Industrial Use of Formal Methods

Industrial Use of Formal Methods

Formal Verification

Edited by
Jean-Louis Boulanger

iSTE

WILEY

First published 2012 in Great Britain and the United States by ISTE Ltd and John Wiley & Sons, Inc.

ISTE Ltd
27-37 St George's Road
London SW19 4EU
UK

www.iste.co.uk

John Wiley & Sons, Inc.
111 River Street
Hoboken, NJ 07030
USA

www.wiley.com

Library of Congress Cataloging-in-Publication Data

Industrial used of formal methods: formal verification / edited by Jean-Louis Boulanger.
p. cm.
Includes bibliographical references and index.
ISBN 978-1-84821-363-0 (hardback)
1. Systems engineering--Data processing. 2. Computer simulation. 3. Formal methods (Computer science) 4. Computer software--Verification. 5. Nondestructive testing. I. Boulanger, Jean-Louis.
TA168.I45 2012
005.101--dc23

2012011854

British Library Cataloguing-in-Publication Data
A CIP record for this book is available from the British Library
ISBN: 978-1-84821-363-0

MIX
Paper from responsible sources
FSC www.fsc.org FSC® C013604

Table of Contents

Introduction

Context

Although formal analysis programming techniques (see works by Hoare [HOA 69] and Dijkstra [DIJ 75]) may be quite old, the introduction of formal methods dates from the 1980s. These techniques enable us to analyze behavior of a software application, described in a programming language. Program correction (good behavior, program stop, etc.) is thus demonstrated through a program proof based on the weakest precondition calculation [DIJ 76].

It took until the end of the 1990s before formal methods (Z [SPI 89], VDM [JON 90]), or the B method [ABR 96, ARA 97] could be implemented in industrial applications, or were usable in an industrial setting.

One of the stumbling blocks was implementing them in the framework of an industrial application (size of application, cost constraints, or delays, etc.). This implementation is only possible using "sufficiently" mature and high-performance tools.

It should be noted that in the context of safety-critical applications, at least two formal methods have a currently-used and recognized environmental design, which covers one part of the specification production process according to the code, all the while integrating one or more verification processes; namely the B method [ABR 96] and the LUSTRE language [HAL 91, ARA 97] and its graphic version, named SCADE[1] [DOR 08].

Introduction written by Jean-Louis BOULANGER.

1 The SCADE development environment is marketed by the company Esterel Technologies; to find out more, visit www.esterel-technologies.com.

The B method and the SCADE environment are associated with industrial tools.

For example, Atelier B and the B-Toolkit, marketed by CLEARSY[2] and B-Core[3] respectively, are tools, which completely cover the development cycle proposed by the B method (specification, refinement, code, and proof generation). Atelier B[4] can be accessed for free from version 4.0 onward.

Formal methods rely on different formal verification techniques such as proofs, *model checking* [BAI 08], and/or simulation.

The use of formal methods while in full development remains marginal, given the number of lines of code. In effect, there are currently many more Ada [ANS 83], C [ISO 99] or C++ lines of code, which have been produced manually rather than through a formal process.

That is why other formal techniques have been implemented to verify the behavior of a software application written in a programming language such as C or Ada. The technical principle known as *abstract interpretation* [COU 00] of programs makes it possible to evaluate all the behaviors of a software application through a static analysis. This type of technique has, in the last few years, given rise to several tools such as Polyspace®[5], Caveat[6], Absint[7], Frama-C[8], and/or ASTREE[9].

The efficacy of these static program analysis techniques has progressed greatly with the increase in power of business machines. It should be noted that these techniques generally necessitate the integration of complementary information such as pre-conditions, invariants and/or post-conditions in the manual code.

SPARK Ada[10] [BAR 03] is one approach where the Ada language [ANS 83] has been expanded to introduce these complementary elements (pre-, post-, invariant), and an adapted suite of tools has been defined.

---

2 To find out more about the CLEARSY company and Atelier B, visit www.clearsy.com.
3 The B-Toolkit was distributed by B-Core (UK) Ltd.
4 Atelier B and associated information can be obtained from www.atelierb.eu/.
5 For more information on Polyspace®, visit www.mathworks.com/polyspace®.
6 To find out more about Caveat, visit www-list.cea.fr/labos/fr/LSL/caveat/index.html.
7 To find out more about Absint, visit www.absint.com.
8 To find out more, visit http://frama-c.com/.
9 To find out more about ASTREE, visit www.astree.ens.fr.
10 The site www.altran-praxis.com/spark.aspx offers further information about SPARK Ada technology.

Objective of this book

In [BOW 95, ARA 97], industrial feedback involving formal techniques can be found, and notably, a report on the B method [ABR 96], the LUSTRE language [HAL 91, ARA 97], and SAO+, the predecessor to SCADE[11] [DOR 08]. Other work such as [MON 00, MON 02, HAD 06] provide a panorama of formal methods with a more scientific point of view.

Given the presentation of the context and of the state of the literature, our objective is to present concrete examples of industrial use of formal techniques.

By formal techniques, we mean the different approaches based on mathematics making it possible to demonstrate that a software application abides by some properties.

Note that the standard use of formal techniques consists of making specification and/or design models, but formal techniques are seen more and more as a possibility of making a verification (static analysis of code, demonstration of abiding by properties, good management of floating-point calculations, etc.).

This work is related to two other books by the same authors published by ISTE and John Wiley & Sons in 2012 [BOU 12c, BOU 12d].

The current book is dedicated to the presentation of different formal techniques, such as the SPARK Ada (Chapter 1) Mr. Telo[12] (Chapter 2), Alta Rica (Chapter 3), Polyspace® (Chapter 4), Escher Verification Studio Perfect Developer (Chapter 5) and the B method (Chapters 6 and 7).

The first book is dedicated to the presentation of different formal methods, such as the B method (Chapters 2 and 3), SCADE (Chapters 6 and 7) MATHLAB/ SIMULINK (Chapters 4 and 5) and Control Build[13] (Chapter 8).

[Bou 12d] involves industrial examples of implementation of formal techniques based on static analysis such as abstract interpretation; examples of the use of ASTREE (Chapters 2 and 3), CAVEAT (Chapter 3), CODEPEER (Chapter 6), Frama-C (Chapters 8 and 7), and Polyspace® (Chapters 4 and 5) tools.

I thank all the manufacturers who have taken the time to redraft and improve upon these chapters.

---

11 It should be noted that SCADE was initially a development environment basing itself on the LUSTRE language and that since version 6, SCADE has become an entirely separate language (the code generator for version 6 takes most of its input from a SCADE model, and not a LUSTRE code).

12 To find out more about MaTeLo, visit www.all4tec.net/index.php/All4tec/mateloproduct.html.

13 To find out more about the ControlBuild tool, visit www.geensoft.com/en/article/controlbuild.

Bibliography

[ABR 96] ABRIAL J.R., *The B-Book – Assigning Programs to Meanings*, Cambridge University Press, Cambridge, August 1996.

[ANS 83] ANSI, Norme ANSI/MIL-STD-1815A-1983, Langage de programmation Ada, 1983.

[ARA 97] ARAGO, "Applications des méthodes formelles au logiciel", *Observatoire français des techniques avancées* (OFTA), vol. 20, Masson, Paris, June 1997.

[BAI 08] BAIER C., KATOEN J.-P., *Principles of Model Checking*, The MIT Press, Cambridge, MA, 2008.

[BAR 03] BARNES J., *High Integrity Software: The SPARK Approach to Safety and Security*, Addison Wesley, Boston, 2003.

[BOU 11] BOULANGER J.-L. (ed.), *Utilisations industrielles des techniques formelles – interprétation abstraite*, Hermes Lavoisier, Paris, 2011.

[BOU 12a] BOULANGER J.-L. (ed.), *Techniques industrielles de modélisation formelle pour le transport*, Hermes Lavoisier, Paris, 2012.

[BOU 12b] BOULANGER J.-L. (ed.), *Mise en oeuvre de la méthode B*, Hermes Lavoisier, Paris, forthcoming 2012.

[BOU 12c] BOULANGER J.-L. (ed.), *Static Analysis of Software: the Abstract Interpretation*, ISTE LTD, London and John Wiley and Sons, New York, 2012.

[BOU 12d] BOULANGER J.-L. (ed.), *Formal Methods, Industrial Use from Model to the Code*, ISTE LTD, London and John Wiley and Sons, New York, 2012.

[BOW 95] BOWEN J.P., HINCHEY H.G., *Applications of Formal Methods*, Prentice Hall, Upper Saddle River, 1995.

[COU 00] COUSOT P., *Interprétation abstraite, Technique et science informatiques*, vol. 19, no. 1-3, p. 155-164, Hermes, Paris, 2000.

[DIJ 75] DIJKSTRA E.W., "Guarded commands, non-determinacy and formal derivation of programs", *Communications of the ACM*, vol.18, no. 8, p.453-457, August 1975.

[DIJ 76] DIJKSTRA E.W., *A Discipline of Programming*, Prentice Hall, Upper Saddle River, 1976.

[DOR 08] DORMOY F.-X., "Scade 6 a model based solution for safety critical software development", *Embedded Real-Time Systems Conference*, 2008.

[HAD 06] HADDAD S., KORDON F., PETRUCCI L. (eds), *Méthodes formelles pour les systèmes répartis et coopératifs*, Hermes Lavoisier, Paris, 2006.

[HAL 91] HALBWACHS N., CASPI P., RAYMOND P., PILAUD D., "The synchronous dataflow programming language Lustre", *Proceedings of the IEEE*, vol. 79, no. 9, p. 1305-1320, September 1991.

[HOA 69] HOARE CAR, "An axiomatic basis for computer programming", *Communications of the ACM*, vol. 12, no. 10, p. 576-580-583, 1969.

[ISO 99] ISO/IEC 9899:1999, Programming languages – C, 1999.

[JON 90] JONES C.B., *Systematic Software Development using VDM*, Prentice Hall International, Upper Saddle River, 1990.

[MON 00] MONIN J.-F., *Introduction aux méthodes formelles*, Hermes, Paris, 2000.

[MON 02] MONIN J.-F., *Understanding Formal Methods*, Springer Verlag, New York, 2002.

[SPI 89] SPIVEY J.-M., *The Z notation – a Reference Manual*, Prentice Hall International, Upper Saddle River, 1989.

Chapter 1

SPARK – A Language and Tool-Set for High-Integrity Software Development

1.1. Introduction

SPARK[1] is used worldwide and across a range of industry sectors for the cost-effective development and assurance of high-integrity software. In this chapter, we will consider the SPARK language and the tools that support its use in increasingly large-scale, high-integrity industrial applications.

The material of this chapter is divided into three main parts, as follows:

– first we present an overview of what SPARK is, including its founding principles, with simple examples;

– next, we provide descriptions of a range of industrial applications of SPARK spanning more than 15 years of use, and supported by some statistics on the size of programs tackled and results achieved, again providing some illustrative material where it is possible to do so;

– we finish with some conclusions drawn from the examples and applications that we have discussed, and a discussion of what is in the pipeline for future enhancements to SPARK and its support tools.

Chapter written by Ian O'NEILL.

1 The SPARK programming language is not sponsored by or affiliated with SPARC International Inc. and is not based on the SPARC architecture.

1.2. An overview of SPARK

1.2.1. *What is SPARK?*

SPARK is a high-level, high-integrity software development language, supported by powerful tools. The compilable elements of SPARK are a subset of Ada, but SPARK is not just a subset of Ada; an integral part of the language is its annotation language, which forms an essential part of the "contractual" specification of SPARK programs.

One element of a program's contract, in both Ada and SPARK is its "signature": its name, arguments (together with their types and modes) and description (in accompanying comments). SPARK programs have additional contractual elements, however. These additional elements consist of a set of core annotations, which may also be supplemented by more specific proof annotations. The core annotations allow a fast, polynomial time analysis of SPARK source code to check for data-flow and information-flow errors, which can indicate a failure of the code to meet its contract. The proof annotations are optional; when these are used, they can support the mathematical proof of properties of the source code. The proofs performed can range from proof of exception freedom, at its simplest, all the way to a proof of correctness against a formal specification. The relationship of SPARK, with its additional annotations, to the "full" Ada language is illustrated in the following Figure 1.1.

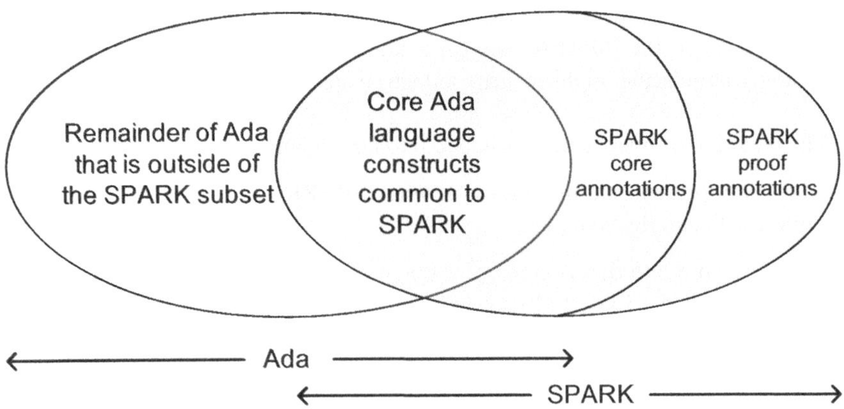

Figure 1.1. *The relationship between full Ada and SPARK*

The subset of Ada, which is at the heart of the SPARK language, has been chosen to produce a simple, yet powerful programming language, retaining the key features that support the construction of software that is demonstrably correct and abstraction through specification. For example, packages, private types, and separate compilation are all important aspects of the language, allowing abstract data types and abstract state machines to be specified, designed, and implemented in a natural way with SPARK. The separate compilation of the package specification and its body mean that developers can design the package specification, which forms the contract for users of the package, separately from the body, and "hide" implementation details from the users of the package, who should not need to see such details to understand what the package does and how to use it. Furthermore, the verification of the body of each package is independent of the bodies of any other packages; only the contracts of the packages on which it depends are needed for verification purposes.

The core SPARK language is a sequential programming language. However, in more recent years, the SPARK language and its principal support tool, the *Examiner*, have been extended to provide support for concurrency in SPARK. This was again achieved by taking a subset of Ada concurrency features that is enforced by the Examiner, supplemented by additional core annotations to express contractual constraints.

The extension of the sequential SPARK programming language with this support for concurrency is called RavenSPARK, which reflects the origins of the SPARK concurrency model in the Ravenscar profile of Ada 95 and Ada 2005.

Since SPARK is designed with high-integrity software development in mind, let's now look at the main influences on the language and how these influences are reflected in the language subset features and the supporting contractual annotations.

1.3. The rationale behind SPARK

For high-integrity software, it is important that the code be unambiguous, amenable to sound and rigorous analysis, and behave predictably when executed. The soundness of the language was therefore a major consideration in the design of SPARK, and in the choice of a subset of Ada in support of this. The original Language Reference Manual for SPARK identified six major considerations in the definition of the language:

– logical soundness;

– simplicity of formal language definition;

– expressive power;

– security;

– verifiability;

– bounded space and time requirements.

[BAR 06a] *discusses three related considerations*: correspondence with Ada, verifiability of compiled code, and simplicity of run-time systems. We shall briefly consider each of these and their implications for the language, before turning to the distinctive aspects of flow analysis and "*Correctness by Construction*" that SPARK makes possible.

Logical soundness

The meaning of the code should be completely and unambiguously determined by its text, and should not be affected by considerations such as elaboration order or order of evaluation of sub-expressions. Thus, an assignment such as:

```
Y := f(X) + g(X)
```

is not legal in SPARK if the order of evaluation affects the result (because f or g may have a *side-effect* on the value of X for instance, or on the value of some other global parameter used by both functions). The Examiner enforces the rules of SPARK, which prevent such order-dependence.

Simplicity of formal language definition

If a language is complex and difficult to define, it will be difficult for users to understand, and more likely for errors of misunderstanding to lead to code that does not behave as such users might reasonably expect. Indeed, experience suggests that parts of a language, which are hard to define precisely are more likely to be a source of compiler problems; conversely, simplicity is likely to make it easier to reason formally about the behavior of the code. Another aspect of this is that the subset of Ada, which is at the core of SPARK typically exercises the most "well-trodden" parts of an Ada compiler, making it less likely that the compiled code suffers from underlying compiler errors.

Expressive power

While simplicity of language definition is to be encouraged, we nevertheless do not want a language, which is so small and constrained that it is impossible to write significant applications in the language. SPARK supports abstraction, hierarchy, modularity, decomposition, low-level and real-time programming, as well as providing facilities for enforcing strong contractual interfaces between units. The proof of the expressive power of SPARK is well reflected in its use in real industrial applications in the numerous lines of code, as we shall see later in this chapter.

Security

By security, we mean that the language lacks insecurities. An insecurity present in Ada, for instance, entails the possibility of aliasing via parameter passing, where the semantics of the sub-program call can in principle depend on the parameter passing mechanism chosen by the compiler. A program that exhibits such dependency is said to be *erroneous*, but no simple mechanism is provided to detect such an insecurity. The SPARK subset of Ada, augmented by its additional annotations and checked using the Examiner, rules out such language insecurities and makes the SPARK code amenable to sound analysis.

Verifiability

For high-integrity applications, we want to have the facilities to prove properties of our software. One aspect of this is language simplicity, in that the semantics of the language are clear and unambiguous. Another aspect is an annotation language, which supports such proof, allowing us to express preconditions, post-conditions, and other invariants in such a way that with appropriate tool support we can perform the necessary mathematical analysis and prove formally that such properties hold. Furthermore, we want this reasoning to be sound: if it is possible to construct such a proof, then we can have confidence that the required property genuinely holds in all cases. This was a major consideration in the design of SPARK.

Bounded space and time requirements

An important application area for SPARK was in high-integrity embedded systems. In such systems, one frequently wants to be able to establish the maximum amount of memory needed, to show this will not exceed system resources in the embedded system. For this reason, SPARK prohibits dynamic storage allocation, and requires that all bounds on types can be determined statically at compile time. It is equally necessary to establish that there are no unintended infinite loops or execution paths, which take too long to complete. While the SPARK language and the Examiner do not provide explicit support for establishing that such memory and timing requirements are met, the rules of the SPARK language subset mean that such reasoning becomes much more tractable in general than for full Ada without any such constraints (N.B. Other tools may be used to investigate worst-case execution time, for instance RapiTime [RAP] and AbsInt aiT [ABS], and to undertake stack analysis, such as AdaCore's GNATStack [ADA]).

Correspondence with Ada

It was a design goal of SPARK that all legal SPARK programs should also be legal Ada programs, and thus it should be possible to compile any SPARK program with any standard Ada compiler. Thus, although SPARK provides extensions to Ada for flow analysis and formal proof, these are in the form of annotations, which are used by the Examiner but appear to an Ada compiler to be just comments, which do not affect compilation. Since all SPARK annotations are effectively comments as far as an Ada compiler is concerned, it is easy to see that SPARK programs are also Ada programs as a consequence of the language definition.

Verifiability of compiled code

For critical applications, it may be necessary to inspect or review the output from the compiler, as an additional verification activity. This may be done to ensure that the executable object code does indeed correspond to the source code that is input to the compiler, and to gain confidence that errors in the compiler have not compromised the correctness of the compiled code. In practice, this can be a cumbersome activity, particularly for a complex language making use of dynamic allocation or constructs, which can make this comparison difficult to undertake. A consequence of the SPARK approach, using a subset of Ada with simpler semantics, static constants and the avoidance of features, which require dynamic allocation and de-allocation is that this activity is somewhat more tractable than it would be for a more complex language.

Simplicity of run-time systems

As already mentioned, an important consideration in designing SPARK was to allow SPARK programs to be used in high-integrity embedded systems. For such systems, we typically require a small run-time system, with certification of the run-time system to meet the required standards for the application area. Again, the design of SPARK is such that a relatively small and straightforward run-time system can typically be used.

1.3.1. *Flow analysis*

With the core annotations, which are an essential part of SPARK, its support tool, the Examiner, can perform various flow analyses to look for potential anomalies in the source code. These are an additional "well-formedness" check, and can be effective in detecting flaws, which might otherwise go unnoticed even after considerable testing.

The flow analyses performed by the Examiner are rigorous and mathematically defined. They are based on those defined and described in [BER 85]. To undertake such analyses with speed and rigor, the source code must be well-structured. SPARK achieves this by a combination of its choice of Ada subset, together with other constraints, which are enforced by the Examiner, such as some restrictions on where exit statements may be placed within loops.

There are two types of flow analysis that the Examiner performs: data-flow analysis and information-flow analysis.

Data flow analysis is concerned with how data flows within and between programs. If we use the value of a variable before it has been assigned a value, for instance, this is a data flow error. Data flow analysis is performed by the Examiner on all of the source code to which it is applied.

Information flow analysis is concerned with how information flows between variables. For a given program, we can determine a set of input (parameters of mode *in* or *in out*, together with globals from which we may read values) and a set of output (parameters of mode *in out* or *out*, together with globals, which we may modify). Information flow analysis considers which input may affect which output, both for individual subprograms and, by composition, subprograms which call these, and so on. As described in [BER 85], information flow analysis can detect more subtle programming errors, including missing or unexpected dependencies between input and output and loop exit conditions, which may become "stable" (i.e. cease to change) after a small number of iterations. Information flow analysis is optional, but it can be applied to all SPARK code, which has the relevant core annotations.

To illustrate the power of the flow analysis performed by the Examiner, consider the following small subprogram, which exchanges its two parameters. We first give its declaration, which includes its signature and its SPARK annotations:

```
procedure Exchange (X, Y : in out Float);
--# derives X from Y &
--#         Y from X;
```

First note that this subprogram has a *dependency-relation* present as a formal annotation (--# comments). This is part of the subprogram's contract. It states that the *exported* value of X is derived (only) from the *imported* value of Y ("derives X from Y"), and that the exported value of Y is derived (only) from the imported value of X. This tells us rather more about the intended behavior of the subprogram than the pure "**in out**" parameter mode information itself.

Now consider a correct implementation of this subprogram:

```
procedure Exchange (X, Y : in out Float)
is
  T : Float;
begin
  T := X;
  X := Y;
  Y := T;
end Exchange;
```

If we apply the Examiner to this source code, we are informed:

```
+++ Flow analysis of subprogram Exchange performed:
no errors found.
```

If, however, we introduce a single coding error, changing the final assignment from "Y := T" to "Y := X" instead, we see a number of data and information flow messages, including:

```
procedure Exchange (X, Y : in out Float)
is
  T : Float;
begin
  T := X;
  ^1
!!! (1) Flow Error : 10: Ineffective statement.
  X := Y;
  Y := X;
end Exchange;
!!! (2) Flow Error : 50: Y is not derived from the
imported value(s) of X.
??? (3) Flow Error :601: Y may be derived from the
imported value(s) of Y.
```

The Flow Analysis chapter of [BAR 06a] covers in detail how these various flow errors and warnings are deduced. Note that just one coding error has given rise to various errors and warnings; let's consider each of the three shown above in a little more detail:

– (1) *Ineffective statement*. The assignment (T := X) is now "ineffective", because T is a local variable, and it is no longer subsequently used to derive a value for any of the subprogram's exports. This is a data flow error, in effect: assigning a value, then failing to make use of it under any circumstances. The remaining two error messages shown are information flow errors;

– (2) *Y is not derived from the imported value(s) of X*. This message tells us that although we've claimed in our contract that the exported value of Y is derived from the imported value of X, this is not what has been implemented. The coding error we've introduced is now discarding the imported value of X without making effective use of it. No matter what the initial value of Y is on entry to this routine, the exported values of both X and Y will be unaffected by it, in other words;

– (3) *Y may be derived from the imported value(s) of Y*. Finally, we find that the exported value of Y now depends on *its* imported value, rather than that of X. This is not what we've claimed in the dependency-relation part of the subprogram's contract, so this discrepancy (between its contract and its behavior) is reported.

Where such "derives" annotations are present and the Examiner is used to perform information flow analysis, any information flow errors or inconsistencies between the contractual specification of the code and its actual behavior will be reported by the Examiner. These derives annotations are not mandatory. However, there is a core annotation, which is mandatory, and which affects the Examiner's flow analysis. Consider adding a global variable counter, which keeps track of how many times exchange gets called, for profiling purposes:

```
...
begin
  T := X;
  X := Y;
  Y := T;
  Counter := Counter + 1; -- added code
end Exchange;
```

If we apply the Examiner to this source code, we are told:+++++++

```
...
Counter := Counter + 1;
^1          ^2
***(1)  Semantic Error  :  1: The identifier Counter is
        either undeclared or not visible at this point.
***(2)  Semantic Error  :  1: The identifier Counter is
        either undeclared or not visible at this point.
end Exchange;
```

We get these error messages because SPARK requires us to state, as part of the subprogram's contract, any global variables that are accessed within the subprogram, together with their mode: *in* (they are only used, not written to), *out* (only written to) or *in out* (both used and written to).

In the case of the exchange sub-program, this means we must also add a global annotation to its specification. In addition, if we are performing information flow analysis, we will need to update its dependency relation to reflect the subprogram's effect on the *counter*. For the correct implementation of the exchange, with its updating of the *counter*, the full specification is now:

```
procedure Exchange (X, Y : in out Float);
--# global in out Counter;
--# derives X from Y &
--#         Y    from X &
--#         Counter from Counter;
```

In short, the *global* and *derives* annotations of SPARK provide a richer, more rigorous description of the code's "contract" with its environment, and make clear to a user of the operation which data values it may read and write, and how these are related, without needing to examine the subprogram body to infer this information. The flow analyses performed by the Examiner both ensure that these contracts are enforced, and detect errors and inconsistencies that may indicate subtle programming errors. Such errors might otherwise go undetected until much later in the development process in languages other than SPARK, by which time they will be much more costly to correct.

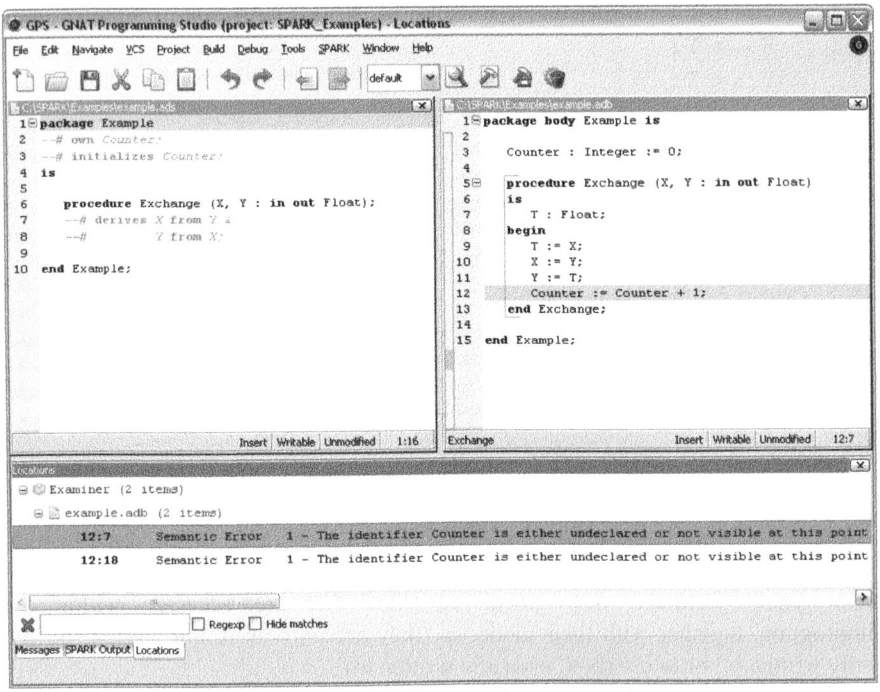

Figure 1.2. *Screen shot from GPS (GNAT Programming Studio)*

Figure 1.2 is a screen shot from GPS (GNAT Programming Studio) showing the final example in this section before correction. The *exchange* procedure has been wrapped inside a package called *example*, the specification for which is shown in the left-hand window and the package body is shown in the right-hand window. The Examiner has just analyzed the package body and the resulting error messages are shown in the window at the bottom. The source code location of the selected error message is highlighted in the right-hand window.

1.3.2. *Code proof*

Along with the *derives* and *global* annotations of SPARK, there is another set of optional annotations known as *proof contexts*. These can be used to express a pre-condition and post-condition on each operation, to provide additional assertions within subprogram bodies (particularly, though not exclusively, within loops), and to declare "*proof functions*" as a mathematical shorthand for use in proof annotations. (Proof functions are not executable functions, in the Ada sense: they are mathematically pure, and can be given a semantics via proof rules, which the user may separately define).

The sort of proof we may wish to perform will vary with the integrity of the application, regulatory requirements, the results of our safety analysis, and so on.

At one extreme, if we have a suitable formal specification of the required behavior of the code, we may attempt formal proof of correctness against this specification. This is the approach that was taken for the most critical SPARK code within the SHOLIS software, discussed here.

At the other extreme, we may not carry out any proof activities at all. In this case, we will gain the benefits of using the SPARK subset of Ada and its stronger contracts imposed via the core annotations and flow analysis.

Aside from carrying out no code proof at all, the most minimal form of proof we may carry out with SPARK is proof of exception-freedom [GAR 93]. This entails proving that the code will not raise a *Constraint_Error* exception at any point during its execution. (A number of other Ada pre-defined exceptions are ruled out by using the SPARK subset while others, such as *Storage_Error*, become more straightforward to reason about as a result of SPARK's boundedness in space and time requirements, and the enforcement of static limits in SPARK).

Such a proof does not show any correspondence between the code and its specification. It merely shows that, whatever it does, it will not use an invalid index into an array, or compute a result outside the range of the relevant base type or subtype, and so on.

Typically, a combination will be used, with proof of exception-freedom for all of the source code, together with targeted proof for critical components. This can involve proving something simpler than pure correctness: for instance, we might instead prove that specific safety invariants are established and maintained.

For example, in an elevator controller, we might want to prove that the elevator movement operations are never invoked with the elevator doors open (via pre-condition constraints), because this is a safety requirement, but we may not wish to prove availability, i.e. that all elevator requests get serviced eventually, because this is not required for the safety case.

Where proof of exception-freedom is being performed, the annotations needed to make the proofs tractable are typically fairly light-weight, provided good use is made of the SPARK language and its type system. For example, for a routine that limits a value to a range *Lower_Bound.. Upper_Bound*, we might see a contract such as:

```
--# pre Lower_Bound <= Upper_Bound;
--# post The_Value in Lower_Bound.. Upper_Bound;
```

The pre-condition states that the range to which we are limiting the value is non-empty, while the post-condition says that the exported value of *The_Value* returned by the routine is within this range. Note, however, that this is quite a "weak" specification of the routine's behavior, as it does not state how the limiting is performed. This may well be all we need, however, for the purposes of proving the absence of run-time errors for this code and for other sub-programs, which make use of it.

If, however, we are performing a more detailed proof then we may need to know more about the operation's effect to prove correct behavior of code that makes use of this operation. We can achieve this by providing a more explicit post-condition, stating in a precise, mathematical style what the subprogram is supposed to do. For example, consider a function ISQRT, which is intended to return the "integer square-root" of a given input:

```
function ISQRT (N : in Natural) return Natural;
--# return Root => Root * Root <= N and
--#               N < (Root + 1) * (Root + 1);
```

As this is a function rather than a procedure, it has a *return* annotation instead of a post-condition. The return annotation tells us how the result returned by the function is related to its input: it may constrain this relationship as little or as much as we require for verification purposes. To read this, note that the "*return Root => predicate*" construct can be read as "return a value *Root* that satisfies *predicate*".

The contract for this function declaration thus says that we want to return the largest Root such that Root*Root <= N still holds (we know it must be the largest, because the second clause says that the square of Root+1 is no longer less than or equal to N).

An important point to note about the contract for ISQRT is that it specifies *what* the function is to return, not *how* the result is to be calculated. It is a mathematical statement of what the function is required to achieve. An implementer is free to choose any appropriate implementation. This could be a simple search, starting at zero and going upward until a suitable value is found, or a more sophisticated binary search, or some other implementation relying on more subtle mathematical properties to determine the result. The key point is that the contract is the specification of what is to be delivered, free from such implementation details.

Thus, SPARK provides the required machinery, in the form of formal annotations, to give a precise mathematical description of the required behavior of the source code, which can form part of the subprogram's contract included in a package specification. Users of the operation can rely on the code meeting this specification provided we prove the mathematical formulae – "verification conditions" (VCs), otherwise called proof obligations – which demonstrate this.

To do so, we first use the Examiner to generate these formulae, which is an entirely mechanical process. Next, we attempt to discharge these proofs, typically using proof tools such as SPARK's Simplifier or other automated proof tools.

Furthermore, note that such proofs are not as difficult or demanding as you might think. They are not of the complexity of establishing the four-color theorem, for instance. In fact, quite the opposite! Proof of exception freedom in particular typically involves proving large numbers of relatively mundane and not-too-demanding formulae, exactly the sort of thing that is amenable to machine-supported proof.

With very little effort, it is common to discharge more than 95% of exception-freedom VCs automatically at an early stage. The remaining small percentage can then be investigated to determine whether they are symptoms of coding errors or not.

In fact, the approach to completing the remaining proofs typically follows a simple iterative approach: the "*proof cycle*", shown in Figure 1.3. In this approach, we annotate with the minimal set necessary, generate VCs, simplify these and then inspect the unproven ones. For those that have not been proven, they may be a symptom of an error, which needs to be rectified, or may just suggest that we need to strengthen or adjust our annotations.

In either case, we fortify to resolve the problem (possibly adjusting our code in doing so), then go around the cycle again: annotate to strengthen or adjust contracts, generate VCs, simplify them, and fortify again as necessary.

More recent experience using this approach suggests that after just a few iterations on the remaining unproved VCs, we can achieve automatic proof of over 98% for well-written code in practice, with very little extra effort. The small number of VCs remaining – typically less than 2% – can then be tackled by other means, including any or all of the following:

– *creation of user-defined proof rules*. The Simplifier allows users to write their own proof rules, which can then be used in attempting to discharge the VCs for particular subprograms;

– *use of other automated proof tools*. As we discuss later, the most recent SPARK Pro toolset provides an interface – SPARKBridge – which allows access to additional provers, such as Alt-Ergo, an SMT solver;

– *use of interactive proof tools*. It is also possible to use interactive proof assistants, such as the SPARK toolset's Proof Checker, to discharge particular VCs, though this is less commonly done;

– *rigorous argument*. The SPARK toolset allows users to record manual reasoning that specific VCs are true. This can be useful on occasion where specific domain knowledge is used, which is outside the scope of the constraints expressed in the source code. In such cases, the developer is responsible for checking the soundness of the reasoning used.

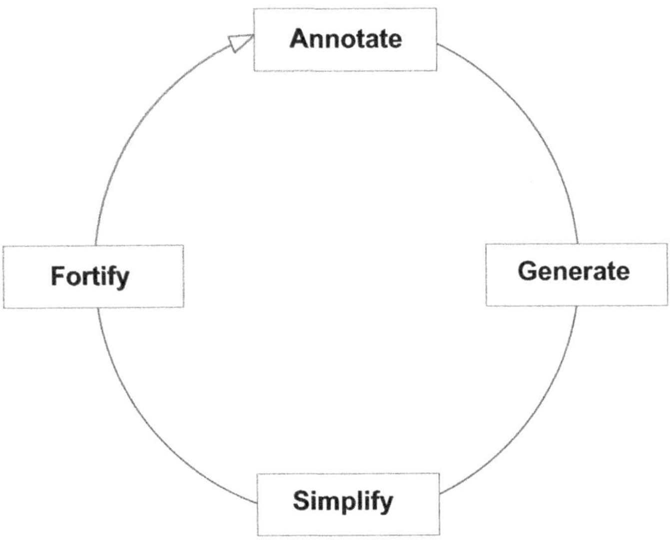

Figure 1.3. *The "proof cycle"*

1.3.3. *Correctness by construction*

An approach that Altran Praxis[2] has applied for many years, and which effectively exploits the language features and design style of SPARK, is "*Correctness by Construction*" (CbyC) [HAL 02, CHA 05]. This approach aims to deliver software with low in-service defect rates at high whole-lifecycle levels of productivity, achieving software, which is robust and resilient to change throughout its lifecycle. The SPARK contractual model provides good support for this.

2 Hereafter, we shall use the simpler name Praxis to refer to the company Altran Praxis Ltd.

Indeed, as reported in [GER 03], the choice of programming language is important in determining a number of factors, including: the level of analysis that is possible; the value achieved by the analyses (strong, rigorous reasoning at one end of the spectrum, versus many false alarms from weaker analysis tools which can stop genuine defects from being detected); and the typical defect density achieved. In Table 1.1 of [GER 03], which compares SPARK with Ada, Pascal, and C among other languages, the average defect density found in the SPARK code is easily an order of magnitude lower than the averages for Pascal and C, six times better than the average for Ada, and an impressive two orders of magnitude better than the worst results for C. Using a sound language and performing analysis at an early stage are major factors in the development of very low defect density software.

CbyC has delivered software with in-service defect rates of 0.4 defects per KLOC or less, yet without compromising productivity. Whole-lifecycle productivity of 30 LOC per day has been achieved with this approach. This compares favorably with the average defect density reported even for the higher CMM Levels 4 and 5 [SOF 09]. Practical experience from a number of projects using CbyC-type approaches for developments ranging in size from 10 KLOC to 200 KLOC confirms that the use of CbyC can consistently lead to low defect software.

SPARK supports the CbyC approach by:

– being a language with *clear and unambiguous semantics*, making it easier to reason about the correctness of the delivered code;

– providing notations for *expressing strong contracts*, making the required behavior of a software component clear to users without the need to delve inside the package body and look at the executable code;

– having tools that enforce these contracts, so that correspondence with the specification can be established with *effective tool support*.

1.4. Industrial applications of SPARK

SPARK has been in industrial use for more than two decades, and there is a growing and active user base today. Long-standing SPARK users include the Eurofighter Consortium (for embedded software on the Eurofighter Typhoon aircraft), Lockheed Martin, Rockwell Collins, Goodrich Engine Controls, the UK Atomic Weapons Establishment, AerMacchi, Westinghouse Rail Systems and ALSTOM Transport, to name but a few.

SPARK applications cover on-board aircraft systems, control systems, rail systems, cryptographic and security applications, in many areas where software integrity is an important consideration.

The following diagram provides a rough timeline of a number of projects and systems in which Praxis have played a part and in which SPARK has been used over the past 15 years (though there is inevitably some overlap).

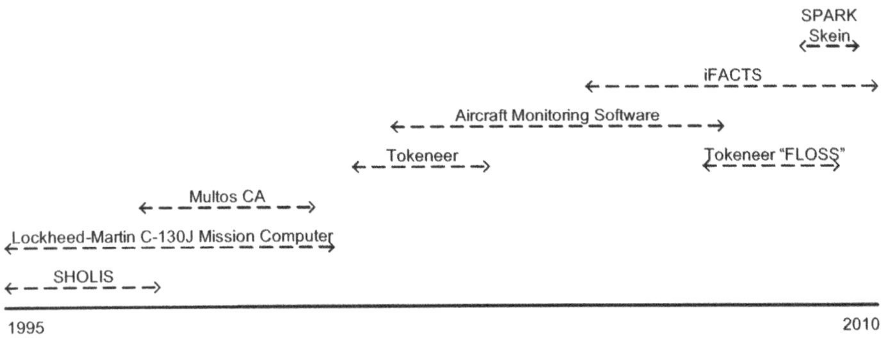

Figure 1.4. *Approximate timeline of applications described in the following sections*

Lack of space and commercial confidentiality prevent us from disclosing a wider range of applications. Nevertheless, in the sections that follow, we will discuss various aspects of each of the applications shown in Figure 1.4, following roughly the order shown on the timeline (from past up to the present day).

1.4.1. *SHOLIS*

SHOLIS, the *Ship/Helicopter Operational Limits Instrumentation System*, is a ship-borne embedded software system that provides advice to a ship's crew on the safety of certain helicopter operations and maneuvers under various scenarios.

At the time of its implementation in the mid-1990s, it was the first such system to be developed to meet the newly introduced rigorous requirements of the UK Ministry of Defense Interim Defense Standard 00-55 for safety-critical software. *Interim Defense Standard* 00-55 had some novel and demanding requirements for the most critical software systems, including:

– the use of a formal specification of system behavior;

– static analysis including the flow analyses of SPARK source code that is performed by the Examiner;

– formal verification (proof) at both the specification and the source code level (with the latter to prove that the code meets the specification);

– independence of the development and verification/validation teams.

Part of the SHOLIS software was assessed to be at the highest *Safety Integrity Level*, SIL 4, requiring the use of all of the above techniques on the relevant components of the SHOLIS software.

Overall, the source code for SHOLIS consists of approximately 27,000 lines of SPARK, of which 14,000 are executable statements (as opposed to declarations). Full data and information flow analysis of all of the source code was performed by the Examiner, and proof of exception-freedom was also carried out for all of the code. The flow analysis performed on SHOLIS supported the demonstration of the separation of the safety-critical and non-critical code elements.

In addition, for the safety-critical code elements, the code was proved to be a correct implementation of the corresponding formal specifications, written in the Z notation [SPI 85], and proof that the code maintains the system's top-level safety properties was also performed.

Proof of the verification conditions generated from the SHOLIS source code entailed proving some 9,000 VCs. Of these, more than 90% were discharged automatically by the Examiner and Simplifier. In the decade or so since the SHOLIS software was first developed, the proof tools have been further improved, so that using the latest version of the Simplifier improves the success rate for automatic proof to more than 95%, more than cutting in half the number of VCs requiring additional manual proof effort.

Development of the SHOLIS software demonstrated that the use of formal methods, proof of correctness and exception-freedom for SPARK code was a real and practical possibility. In addition, the SHOLIS development project revealed that proof of both the Z formal specification and the SPARK source code can be significantly more cost-effective at finding faults than traditional testing activities [KIN 00], in part because they allow faults to be found and corrected earlier in the development process.

1.4.2. *Lockheed-Martin C-130J mission computer*

> "Very few errors have been found in the software during the most rigorous levels of FAA testing, which is being successfully conducted for less than a fifth of the normal cost in industry".
>
> — Jim Sutton, Lockheed Martin

Figure 1.5. *A Quote*

The Lockheed-Martin C-130 transport aircraft, commonly known as the "*Hercules*", was updated in the 1990s to produce the C-130J Hercules II, with significantly improved avionics, engines, and propellers.

Part of the transformation involved new *Operational Flight Plan* (OFP) software [CRO 96, MID 05] to control the many individual systems on the aircraft and to operate and provide information on the integrated aircraft. The primary OFP consists of two mission computers, with backup OFPs in each of the two *Bus Interface Units*.

The C-130J OFP software was required to be certified for both civil and military use. This meant meeting both DO-178B for civil applications, as well as the static analysis and formal verification requirements of the UK's *Royal Air Force*.

The OFP software consists of around 200 KLOC of source code, written in Ada. Of this, all components with significant system semantics are implemented in SPARK, allowing proof of correctness against a specification expressed using a notation known as CoRE, or "*Parnas tables*".

These tables are written so that each line of the table specifies an element of the overall input-to-output function required. The union of the lines of the table should cover all possible input value combinations, while there should be no overlap between distinct lines of the table (unless the overlap is intended, and yields the same output value in all cases of such overlap).

This form of specification is well-suited to analysis with SPARK. The overall flow from input to output, illustrated below, maps very well on to SPARK dependency relations ("*derives Output from* $Input_1$*, ... ,* $Input_n$", in effect). Each line of the Parnas table can be expressed, suitably formalized, as a clause within the corresponding post-condition of the relevant operation, allowing us to prove correctness with respect to the specification. Furthermore, any issues or anomalies encountered in formalizing the Parnas table as a post-condition are raised for investigation and resolution, thereby detecting errors and inconsistencies earlier in the development lifecycle than might otherwise be the case.

An example of a Parnas table is shown below (note that "X" entries mean "don't care"), together with how this would typically be represented as a SPARK post-condition.

Tank_Level	**Xfer_Method**	**Direction**	**Control_Valve**
"X"	"X"	Empty | Off	Shut
"X"	Manual	Fill	Open
< Limit	Auto	Fill	Open
>= Limit	Auto	Fill	Shut

Table 1.1. *Example Parnas Table*

```
--# post
--#  ((Direction  =  Empty  or  Direction  =  Off)  ->
Control_Valve = Shut)
--# and
--# ((Xfer_Method = Manual and Direction = Fill) ->
Control_Valve = Open)
--# and
--# ((Tank_Level < Limit and Xfer_Method = Auto and
Direction = Fill) ->
--#          Control_Valve = Open)
--# and
--#  ((Tank_Level  >=  Limit  and  Xfer_Method  =  Auto
and Direction = Fill) ->
--#         Control_Valve = Shut);
```

The use of SPARK proved to be a major success, with [CRO 96, MID 05] reporting that the use of formal methods was "advantageous from an economic standpoint", while "the technology for generating and discharging the proof obligations, based on the formal definition of the SPARK component of Ada, was crucial, in binding the code to the initial requirements". The use of SPARK contributed directly to the economic savings reported, in helping to produce code with an unusually low defect density.

1.4.3. *MULTOS CA*

As reported in [HAL 02, CHA 00], the *Multi-Application Operating System* (MULTOS) is an operating system designed for use on smart cards. It allows several applications to reside on a single such device and enforces separation of applications. Digital certificates signed by the MULTOS Certification Authority (CA) are used to prevent forging of smartcards and applications, which is a key security concern.

The system at the heart of the MULTOS CA, which issues these digital certificates, is subject to stringent security constraints and is required to meet the standards of level "E6", the highest assurance level of the UK ITSEC scheme [ITS 91].

A requirement of the software for the system was that it should be designed to run uninterrupted. The software cannot be upgraded or even restarted without significant effort. As a result, avoiding memory leaks or unexpected behavior, such as termination via an exception, was a major goal. The system components were furthermore classified into a number of security categories, with the user-interface and data coming from it considered insecure, for instance, and requiring rigorous validation.

The security-enforcing core of the system was implemented in SPARK. SPARK flow analysis made an important contribution to establishing the absence of subtle security vulnerabilities: failure to initialize a variable might lead to unencrypted information leaking out of the system, for instance, so data flow analysis can be used to demonstrate the absence of this particular vulnerability. Information flow analysis was also useful in this respect, showing that certain data elements cannot affect other elements, giving confidence that such security properties were being established and maintained by the code.

The development of the MULTOS CA software demonstrated that SPARK can aid in showing that the security requirements and separation of concerns are achieved.

Furthermore, despite the stringent analysis and verification activities, good productivity of 28 LOC per day was achieved, even when measured over the whole development lifecycle. Finally, the overall defect density achieved was also very low: 0.04 defects per KLOC were found in service one year after delivery, in an overall system of roughly 100 KLOC.

1.4.4. *Tokeneer*

To study the effectiveness of SPARK and the CbyC approach, and its relevance to the security community, the US National Security Agency (NSA) commissioned Praxis to redevelop the software for the Tokeneer ID Station (TIS) using SPARK and CbyC [BAR 06b] (the Tokeneer ID Station is one component of a larger system, Tokeneer, which was originally developed by the NSA as a research vehicle for investigating various aspects of biometrics as applied to access control).

Tokeneer itself consists of a secure enclave, physical access to which must be controlled, and within which are a number of workstations. To gain access to the enclave and to the workstation, users have security tokens (for instance, smart cards), which must be presented to a reader external to the enclave. The reader uses information encoded on the entry token to carry out biometric tests (such as fingerprint reading). The door to the enclave will only be opened if the user passes these tests, allowing the user to enter the enclave and use the token to gain access to the workstations.

At the heart of the TIS is the TIS Kernel, on which the security assessments and certifications were to focus. The TIS Kernel, intended for use in high risk environments, was required to meet evaluation assurance level (EAL) 5 of the Common Criteria [ISO 99].

To summarize, the main steps of the development process used to meet the TIS Kernel requirements were as follows:

– requirements analysis. The most expensive mistakes to fix are those that occur early on in the development process. The Praxis team used the REVEAL[3] requirements engineering approach to investigate and clarify the system boundary, which also helped to clarify dependencies on aspects of the environment outside the TIS Kernel, such as certificates and the behavior of the door and its latch;

– security analysis. This involved creating a Security Target and a Security Policy Model, which seek to identify only the key properties that must be satisfied to ensure security, and ignore other aspects such as functionality, ease of use, etc. This also helps in gaining further insight into the requirements, and can be used to prove the functional specification has the required security properties, thereby acting as a form of validation of the specification;

– specification. CbyC requires that we develop a sufficiently detailed, rigorous specification before beginning to write code. It is an essential step in documenting what you expect the system to do, rather than leaving it to the coders to guess this. It also provides a frame of reference for others in the development team. When the code is designed and written, design and code review activities will be more fruitful and rigorous if the review is focused on checking that the code delivers the functionality documented in the specification. In addition, testers will be able to derive tests from the specification, to test that the code does indeed meet the specification. For the TIS Kernel specification, the Z notation was used;

– design. Prior to writing the executable code, one or more design steps are applied, depending on the size of the system and the complexity of the design decisions to be made. For the TIS Kernel, two further documents were developed:

- a refinement (again in Z) of the functional specification,

- an outline of the SPARK packages to be developed, including the location of a persistent state, using notations from the INFORMED design method for SPARK [INF 05].

These documents had clear purposes, including the documentation of difficult implementation decisions, the detailed definition of the structure and content of certain audit elements and how these should be stored, the definition of the concrete representation of certificates, the provision of a top-level design of the main software components and the information flow between them, which can be analyzed with the Examiner:

– *implementation*. The package structure outlined in the design stage can then be fleshed out with SPARK package bodies implementing the required functionality,

3 REVEAL is a UK registered trademark of Altran Praxis Ltd.

using the Examiner early and often to demonstrate consistency and well-formedness from an early stage of implementation;

– *system test*. The approach taken was to produce incremental builds throughout development, with progressively more functionality in each build. This allowed integration problems, which tend to be the most problematic, to be tackled and tested early. It also removed the need for a separate expensive integration phase. Testing was conducted against the refined functional specification produced during design. Because the specification is rigorous, testing against it is much more effective, since the specification covers *all* system behavior clearly, including error conditions and special cases as well as correct behavior;

– *assurance*. Every step in the development process and every decision made has the potential to introduce errors. However, because of the precision and rigor of the documents and code produced at each stage, it is possible to undertake meaningful analysis to attempt to identify such errors as early as possible. Figure 1.6 illustrates the assurance activities undertaken throughout the Tokeneer development, with arrows showing the input to each assurance activity, and arrows out of an assurance activity showing which lifecycle product is being validated.

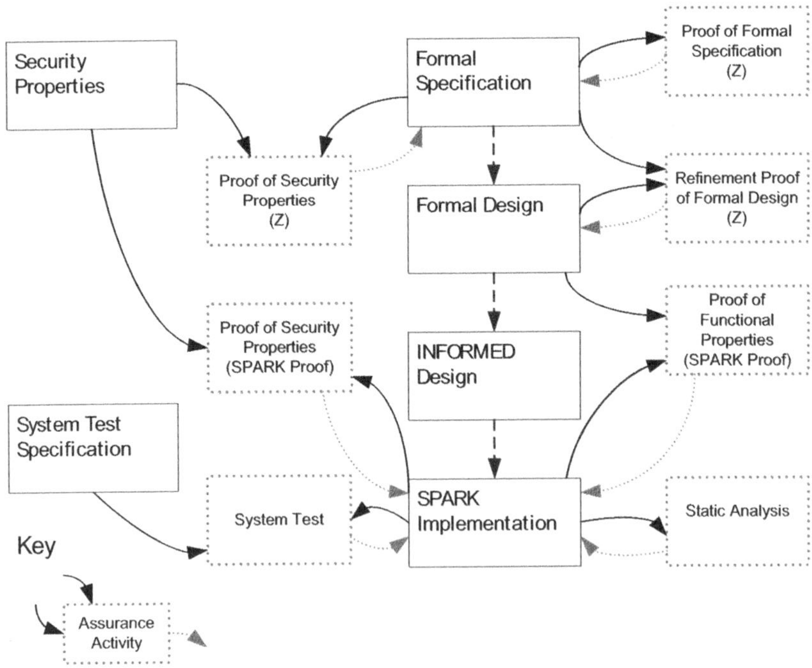

Figure 1.6. *This shows the main assurance activities undertaken in the TIS Kernel software development. Note that assurance of the INFORMED design is achieved by analyzing with the Examiner from an early stage of development*

The results achieved were once again impressive. Very few defects were found post-development in the 10 KLOC of source code, which was developed at an overall rate of 38 LOC per day. The Tokeneer project materials have been made available publicly by the NSA, including the complete set of development documentation and source code for research and study purposes (see for example http://www.adacore.com/tokeneer). Even with the rigorous scrutiny of independent experts and academic researchers, very few defects have been identified post-delivery: [WOO 10] reviews those that have, including via the use of more recent sophisticated analysis tools for Ada.

1.4.5. *Aircraft monitoring software*

This is another embedded aircraft system, the design approach for which is described in [AME 04], with software written in SPARK using the CbyC approach.

A novel aspect of the approach was the use of UML, with INFORMED [INF 05] to design the structure of the software components and guide the construction of the UML model. The overall design approach entailed:

– the use of INFORMED to construct a design for the relevant system component;

– the capture of the design in UML, including component contracts (in the form of SPARK annotations);

– the validation of the system design at an early stage through the use of the Examiner to check for errors;

– the implementation in SPARK, using the Examiner to establish that the code meets its contract.

UML stereotypes were used to define new elements, such as a SPARK Ada package. UML-tagged values were used to include essential SPARK elements within the UML model, such as SPARK core annotations for packages and sub-programs.

This allowed the UML model to be constructed in such a way that the overall *structure* of the SPARK source code could be generated from the UML model. This provided the main objects – package, type and sub-program definitions – of the SPARK source code, but not the executable sub-program bodies: these were hand coded, and the Examiner was used to ensure that the hand-written code for each unit met its contract, with proof of exception-freedom of the source code performed as a standard part of the development process.

1.4.6. *iFACTS*

NATS, the UK's leading air traffic services provider, has pioneered research and development into new predictive (look-ahead) tools for air traffic controllers. iFACTS – the *interim Future Area Controls Tools Support* – provides tools for trajectory prediction, conflict detection, and monitoring aids. It will replace traditional paper information strips with electronic data and new displays. iFACTS enables air traffic controllers to increase the amount of air traffic they can handle, providing capacity increases and enhancing safety.

NATS appointed Praxis to develop the new software for iFACTS, which needed to meet SW01, the stringent software safety regulation required by the UK Civil Aviation Authority. Praxis applied its Correctness by Construction method, deployed a multi-disciplinary, multi-national team of over 100 engineers, and worked with NATS at all levels to ensure success.

The iFACTS source code consists of over 200 KLOC of RavenSPARK source code, from which over 120,000 verification conditions are generated to prove exception-freedom. The Examiner and the Simplifier between them prove some 98.5% of these VCs entirely automatically, without the aid of any user-defined proof rules.

All of these VCs can typically be generated and simplified automatically overnight on a single, reasonably powerful desktop PC. Of the remaining approximately 1,800 VCs, some 1,500 are also discharged automatically once user-defined proof rules are used; again, this can be achieved as part of the overnight run of the SPARK tools on the overall source code. This has been carried out throughout development, allowing any flow errors or unproven VCs resulting from changes to the source code to be detected and addressed as soon as possible.

Initial versions of iFACTS have successfully been deployed to the live operational environment at Swanwick's en-route center and used for shadowing and operational trials at quiet times of the day. It became fully operational in 2011.

1.4.7. *SPARK Skein*

One perceived weakness of SPARK, and Ada in general, is that it entails greater overheads than a rather raw, lower-level language such as C, which provides minimal protection against coding errors and minimal run-time overheads. This perception is, however, entirely incorrect. It is possible to take full advantage of the benefits of SPARK and still achieve performance that is the same as that which can be achieved with less supportive languages, as a recent paper [CHA 10] on the development of SPARKSkein has shown.

Skein is a new family of cryptographic hash functions, designed to be fast, secure, and simple. The reference implementation on [SKE] is in C. As reported in [CHA], Praxis and AdaCore engineers re-engineered this implementation into a reference SPARK implementation, identifying a code flaw in the reference C implementation in the process. Through the use of appropriate compiler switches with the GNAT compiler used for this project, it was possible to achieve similar levels of efficiency in the SPARK executable code, indeed performance which could surpass that of the comparable C implementation in the right circumstances.

The SPARKSkein implementation project demonstrates not only that it is possible to have rigor without loss of efficiency, but also that SPARK is a powerful and flexible development tool applicable to a wide variety of application areas.

1.5. Conclusion

The SPARK language provides a high-integrity language for the development of software for critical applications, supported by powerful, industrial-strength analysis tools such as the Examiner and Simplifier. Furthermore, as the case studies we have looked at have consistently shown, the use of a rigorous language and powerful analyses can yield lower defect software without any reduction in whole-lifecycle productivity.

Indeed, we would argue that the benefits that have been demonstrated are so great that the use of SPARK in any project, whether critical or not, makes good economic sense. This is a reflection of the fact that using such techniques, it is possible to trap errors earlier in the development process, leading to higher-quality code by the time that testing commences and less time spent in testing.

Recent developments in SPARK include the launch of the GPL (*General Public License*) version of SPARK, which aims to drive the use of SPARK high-assurance programming techniques and tools by a larger proportion of the overall software development community. This has been undertaken in partnership between Praxis and AdaCore. The professional toolkit – SPARK Pro – is now globally available from AdaCore.

The more recent SPARK Pro toolset offers additional tools not present in earlier releases of the SPARK tools. This includes "ZombieScope", which allows the Examiner and the Simplifier to be combined to search for non-executable or "dead" paths through the source code (hence the name). It is also available on a wider range of platforms and supports extensions such as the automated checking of safety and security policies through flow analysis.

Feedback from the SPARK user community continues to drive research and development of the SPARK language and toolset. Extensions already in the pipeline or planned for the near future include the following:

– extensions to the SPARK language are under development at the time of writing, including support for the use of Ada generics with appropriate subset rules, together with provision of a standard container library for SPARK;

– SPARK Pro release 9.1 provides SPARKBridge, the first interface to additional proof tools other than the simplifier. At present, this is being used successfully with Alt-Ergo, as reported in section 1.4 above, with the possibility of using it with other SMT provers;

– integration of SPARK with model-based design tools such as SCADE.

For the latest details of the SPARK language and supporting tools, plus additional background technical papers, please visit http://www.altran-praxis.com/spark.aspx and http://www.adacore.com/home/products/sparkpro.

1.6. Bibliography

[ABS] See http://www.absint.de/ait.

[ADA] See http://www.adacore.com/home/products/add-on_technologies/stack_analysis.

[AME 04] AMEY P., WHITE N., "High-integrity Ada in a UML and C world", *Lecture Notes in Computer Science 3063, Reliable Software Technologies – Ada-Europe 2004*, by LLAMOSÍ A., STROHMEIER A. (eds), p. 225-236, http://www.altran-praxis.com/downloads/SPARK/technicalReferences/ada_uml_and_c.pdf, Springer-Verlag, New York, 2004.

[BAR 06a] BARNES J., *High Integrity Software: The SPARK Approach to Safety and Security*, Addison-Wesley, Boston, 2006.

[BAR 06b] BARNES J., CHAPMAN R., JOHNSON R., WIDMAIER J., COOPER D., EVERETT W., "Engineering the tokeneer enclave protection software", *1st IEEE International Symposium on Secure Software Engineering*, http://www.altran-praxis.com/downloads/SPARK/technicalReferences/issse2006tokeneer.pdf, March 2006.

[BER 85] BERGERETTI J.-F., CARRÉ B.A., "Information-flow and data-flow analysis of while-programs", *ACM Transactions on Programming Languages and Systems*, ACM, New York, vol. 7, p. 37-61, January 1985.

[CHA 00] CHAPMAN R., "Industrial experience with SPARK", *ACM SigAda Conference*, http://www.altran-praxis.com/downloads/SPARK/technicalReferences/ind_experience_spark.pdf, 2000.

[CHA 05] CHAPMAN R., "Correctness by construction: a manifesto for high-integrity software", *Proceedings of the Tenth Australian Workshop on Safety-Related Programmable Systems (SCS 2005), CRPIT*, vol. 55, p. 43-46, ACS, Sydney, Australia, 2005.

[CHA 10] CHAPMAN R., BOTCAZOU E., "SPARKSkein: a formal and fast reference implementation of skein", http://www.skein-hash.info/sites/default/files/SPARKSkein.pdf, 2010.

[CRO 96] CROXFORD M., SUTTON J., "Breaking through the V and V Bottleneck, presented at Ada in Europe 1995", *Lecture Notes in Computer Science*, vol. 1031, http://www.altran-praxis.com/downloads/SPARK/technical References/v_and_v_bottleneck.pdf, Springer-Verlag, 1996.

[GAR 93] GARNSWORTHY J.R., O'NEILL I.M., CARRÉ B.A., "Automatic proof of absence of run-time errors", *Proceedings of Ada UK Conference*, London Docklands, October 1993.

[GER 03] GERMAN A., "Software static code analysis lessons learned, crosstalk", http://www.stsc.hill.af.mil/crosstalk/2003/11/0311German.html, November 2003.

[HAL 02] HALL A., CHAPMAN R., "Correctness by construction: developing a commercial secure system", *IEEE Software*, p. 18-25, Jan-Feb 2002.

[INF 05] INFORMED, Design method for SPARK, S.P0468.42.4, Altran Praxis Ltd., January 2005.

[ISO 99] ISO 15408, Common Criteria for Information Technology Security Evaluation, version 2.1, August 1999.

[ITS 91] ITSEC, Information Technology Security Evaluation Criteria, Provisional Harmonised Criteria, version 1.2, June 1991.

[KIN 00] KING S., HAMMOND J., CHAPMAN R.C., PRYOR A., "Is proof more cost effective than testing?", *IEEE Transactions on Software Engineering*, August 2000.

[MID 05] MIDDLETON P., SUTTON J., *Lean Software Strategies*, Productivity Press, New York, 2005.

[RAP] See http://www.rapitasystems.com/rapitime.

[SKE] See http://www.skein-hash.info/

[SOF 09] Software Engineering Institute, "Implementing CMMI for High Performance", *CMMI Made Practical Conference*, http://www.cmminews.com/2009/pdfs-sessions/73.pdf, Carnegie Mellon University, London, April 2009.

[SPI 85] SPIVEY J.M., *The Z Notation: A Reference Manual*, 2nd Edition, Prentice-Hall, Upper Saddle River, 1985.

[WOO 10] WOODCOCK J., AYDAL E.G., CHAPMAN R., "The tokeneer experiments", *Reflections on the Work of C.A.R. Hoare*, JONES C.B., ROSCOE A.W., WOOD K.R. (eds), Springer, New York, 2010.

Chapter 2

Model-Based Testing Automatic Generation of Test Cases Using the Markov Chain Model

2.1. Preliminaries on the test process

2.1.1. *Findings*

Test phases are the last stages of system development. They are therefore severely constrained by delivery periods. In general, they are subject to all the delays accumulated during the overall project.

Currently, test phases still correspond to 30% or even to 50% of the total cost of system development [JON 08] and thus represent an important point for possible improvements.

Testing is a difficult activity. Indeed, it is not possible to test everything because of the sheer numbers for the majority of modern systems. The difficulty of testing lies in the selection of relevant test cases for each phase of validation.

It also lies in the absence of a true reflection of test optimization guided by both quality and costs.

Chapter written by Héléne LE GUEN, Frederique VALLÉE and Anthony FAUCOGNEY.

2.1.2. *Test optimization*

The concept of reliability is based not on the number of remaining faults in the software, but on the probability of the appearance of errors in an operational situation. It is this notion of reliability, which makes it possible to guarantee a time of correct operation in a particular environment. The concept of an operational profile [MUS 99], called a usage model, makes it possible to define the various types of software use.

Since the reliability of the software depends not on the number of residual code defects but on their capacity to be activated in a given usage profile, the various studies carried out around these problems tend to show that the best testing strategy to increase reliability is that of the statistical usage test [VAL 00].

The development methodology Cleanroom [DYE 92] laid the basis of the statistical usage test. Cleanroom technology brought a change to software development and testing while aiming to produce software with close to a *zero-defect* quality.

Its principal characteristic is that it recommends statistical control throughout the lifecycle. Cleanroom does not provide a formal methodology, but it proposes applicable principles to improve development. One of these principles consists of accounting for all of the faults found by inspecting all of the stages of the development process to better control the project.

The tests in Cleanroom also present an innovation: they are there to guarantee a MTTF (*Mean Time To Failure*) of the software. Indeed, it is not a question here to know if everything was tested, but to know the future behavior of the software to guarantee an average time of trouble-free software operation.

2.1.3. *The statistical usage test*

The basis of the statistical usage test therefore results from two previously evoked reports, namely:

– the exhaustive test is not possible in the majority of the situations;

– the frequencies of appearance of certain errors are very weak.

The purpose of the statistical usage test is not to find all of the software's defects, but rather to identify all of the errors having a probability of occurrence higher than a certain threshold. The test is thus carried out in accordance with the envisaged operational use.

Several statistical test techniques have been developed, such as selecting entries according to a random profile [NTA 98], using partitioning [OST 88], modeling a usage profile with a Markov chain [WHI 92], or even automatic derivation of statistical tests starting from formal specifications [THE 93].

Different methods exist for implementing the statistical usage test. Constructing a model representing the software's dynamic behavior and use makes it possible to derive conclusions automatically from the test cases. These tests being automatically derived will be neither intuitive nor similar to the tests written by a person accustomed to drafting test plans. The usage tests greatly put the system to the test because they test a large number of possible usage cases and not just one unit of functionalities.

The statistical test presents some disadvantages. The first is the construction of the usage model. In many of the examples cited in literature, it is often found that it is either simplified system applications (synchronized, Boolean data exchanges, etc.) or applications very dependent on the tested system, in particular with the use of an Oracle, that determine the test's verdict.

The usage model must naturally be able to adapt to a broad range of applications. Attributing probabilities is not easy insofar as each system user could behave differently. However, even a coarse attribution always gives more interesting results than the equiprobability of functionalities.

Employing the statistical usage test, however, has many advantages:

– the most frequently used functions are tested much more often;

– it relates the only method that allows obtaining an estimation of the time of the application's proper function;

– it quickly discovers the most frequently occurring errors and therefore considerably improves the system user's view;

– it becomes rather unlikely, if the model is correct, that an error is discovered in the first moments of system use because the simplest cases of usage are validated.

Remember also that the statistical usage test, like any other validation test strategy, will appear optimal only if the system reaches the validation stage with a reasonable rate of residual faults. Indeed, in the contrary case, it is more interesting to resume an earlier test strategy, unit or integration tests, or even to completely resume a poorly conceived code.

2.1.4. *Generating test cases*

The manual generation of test cases, which is still the most often employed, consists of manually writing the test sequences starting from a reference (for example, the specification). The techniques for choosing these test cases vary. Owing to the essential human element, the test cases may be poorly chosen (from a memory lapse or, on the contrary, from redundantly checking). In addition, with systems that are becoming increasingly complex, it is extremely expensive to implement.

Today, the competition is such that the specifications evolve systematically even during development. Therefore, the maintenance costs of manual plans explode since they require quasi-systematic recoveries.

Automatically generating test cases solves the problem of poorly chosen test sets, so at least the process of generation is efficient and adapted to the system. However it requires a formalized expression for the reference.

NOTE 2.1.– The test sequences generated can, at the same time, comprise the input and the output of the implementation being tested. Whereas the input selection is easy, the generation of the anticipated data (thus of the output produced by the system) is much less commonplace. This difficulty is known as the "Oracle problem". The challenge is to calculate the expected results without having to develop a reference model comparable to the system being tested in terms of size, complexity, and thus cost. However, some solutions do exist, such as the use of semi-formal or formal specifications.

2.2. Modeling using Markov chains

2.2.1. *Origin*

The principles employed to make a test model based on Markov chains depend on the work carried out during the statistical usage test [WHI 92, MUS 99].

Excluding notable exceptions such as the telephone company AT&T, note that these techniques took a long time to become employed by industries.

This is probably due to a lack of practical and easily comprehensible tools to implement them. Specifically, this is what pushed the company ALL4TEC in 2000 to invest in the development of a tool, such as MaTeLo© (*Markov Test Logic*).

At the beginning of the development of the MaTeLo tool, the problems to be solved included adapting statistical approaches so that the industries could gain some of the benefits, i.e. testing with better quality and, if possible, at a lower cost.

This work involves creating a model that more effectively takes into account the systems' various characteristics including the possibility of indicators related to the time of the test remaining to reach a given quality.

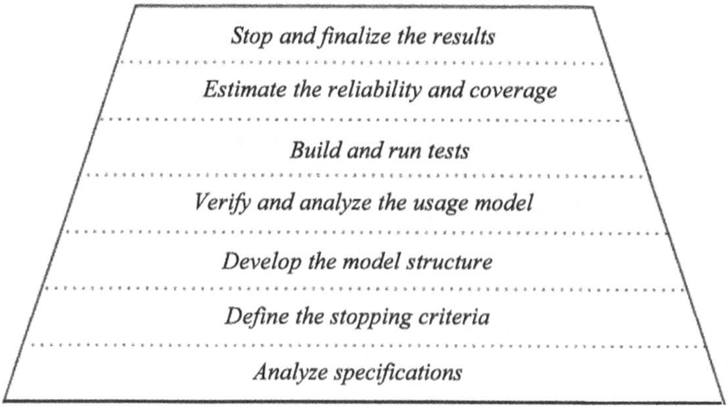

Figure 2.1. *The main stages of the statistical usage test*

The essential components of the method include:

– constructing a model representative of the possible scenarios of system use: the usage model;

– estimating the test load to reach the stopping criteria;

– using the generation algorithms of test cases;

– determining the functional coverage;

– calculating the reliability;

– monitoring the test campaign.

The major interests in this method are:

– the possibility of making a control of the model (*model-checking*) before using it for generation;

– the possibility of guaranteeing the functional coverage by using a formal description of the exchanges between the environment and the system, and then using this model to generate the test case;

– the possibility of measuring reliability after the test's execution;

– reducing the validation time for a quality-controlled delivery;

– maintaining a usage model, which is easier than the maintaining test plans.

This statistical method is based on the construction of a test model in line with the system specification, which represents the complex exchanges between the system being tested and its environment.

2.2.2. *Mathematical formalization*

There are many models for case test generation, which are all based on state machines [PET 00]. The model retained for MaTeLo is a combination of state machine models and Markov chains. Its objective is to propose solutions for conducting realistic test campaigns aiming to create stop test criteria (such as guaranteeing reliability and a functional coverage) for a controlled test's time.

This model uses only one test component, which appears sufficient in a large majority of the cases. Several ports of communication can be declared, and input and output vectors are attached to these ports. Thereafter, we will reconsider the description of the input and output vectors.

The test component models the executive test. The exchanges between the executive test and the system being tested are modeled by a state diagram, which is presented in the following section.

2.2.2.1. *State diagrams*

The usage model consists of a group of state diagrams. The states of the diagrams are stable. A state is described as stable if the system cannot evolve without receiving a new event from the outside (such as an input coming from the executive test).

A diagram always contains two particular states, the {Start} state, which represents the initial state of the system before starting a test and the {End} state, which represents the final state of the system after the execution of a test.

The *stimuli* applied to the system as well as the reactions of the latter are associated with the state machine transitions. A transition is connected to an exchange between the environment and the system being tested, i.e. an input and possibly one or more outputs. This is known as atomic transition, i.e. when a transition on the graph is taken, and then all the actions connected to it will run until the system is in a stable state. This property is similar to Mealy machines.

The input field is, in general, infinite. On the other hand, the number of different types of events that make the system change from a stable state is finite.

An exchange starts with an input (message from the environment to the system being tested) and finishes when the system is in a stable state, i.e. when the system cannot evolve any further without receiving another input.

DEFINITION 2.1.– *The test graph* G *is defined by* G = {E, Start, End, T, AI, AO, d} *such that*:

– E is a finite number of states;

– Start $\in$ E is the initial state of the system; this state is not connected to an incoming transition;

– End $\in$ E is the final state of the system; this state is not connected to an outgoing transition;

– $T = \{t_1, \ldots t_M\}$ is the set of transitions;

– $A_I = \{i_1, \ldots i_{Ni}\}$ is the set of input vectors;

– $A_O = \{o_1, \ldots o_{No}\}$ is the set of output vectors;

– δ (e, i) *is the transfer function that makes it possible to change from a stable state in the test graph by following the transition* $t \in T$: $E \times A_I \rightarrow (o_1, o_2, .. o_n) \in A_O \times E$, *the transition* $t \in T$ *is associated with an input vector* $i_m \in A_I$ *and a set of output vectors* $(o_1, o_2, .., o_n) \in A_O$.

NOTE 2.2.– It is possible according to this definition to have several transitions connecting the same set of states.

NOTE 2.3.– It is not possible to reach the {Start} state beginning at any state of the graph, and the {End} state does not allow any state of the graph to be reached, it acts as a final state.

2.2.2.2. *Typology of events*

The input (*stimuli*) and the output (*reactions*) are the observable events, which can be exchanged between the system and the environment. An input (or an output) is described using typed variables.

We present here the language used to describe the input and output vectors that are associated with the test graph. The events are listed and described only once. In the test graph defined previously, it is then possible to link the same event several times to a transition as an input or output. To then link the state machine to a test architecture, it is necessary to indicate the port associated with each event.

Let $V = (x_1, x_2, \ldots x_H)$ be the ordered set of variables associated with a transition. An input vector $va = (x_{1,a}, x_{2,a}, \ldots x_{H,a})$ is defined by an instantiation of each one of these variables.

The variables x_h can be from eight different types: integer, real, Boolean, object identifier, bits chain, bytes chain, list of elements, or character chain.

It is also possible to define types made up of key words:

– *sequence*, which indicates a structured element made up of several variables of different types;

– *sequence of* which indicates a structured element made up of several variables of the same type;

– *choice*, which makes it possible to define a variable whose type belongs to a list of types.

This classification of events makes it possible to define the input and output vectors in a more precise way. An input vector will be described according to a set of instantiated variables.

The model is built so that all the input vectors associated with the same transition are defined according to the same set of variables.

2.2.2.3. *Hierarchical representation*

The more complex a system is, the more difficult it is to model. It is therefore necessary to have the possibility of introducing hierarchical levels, which allow a certain abstraction in the model's description.

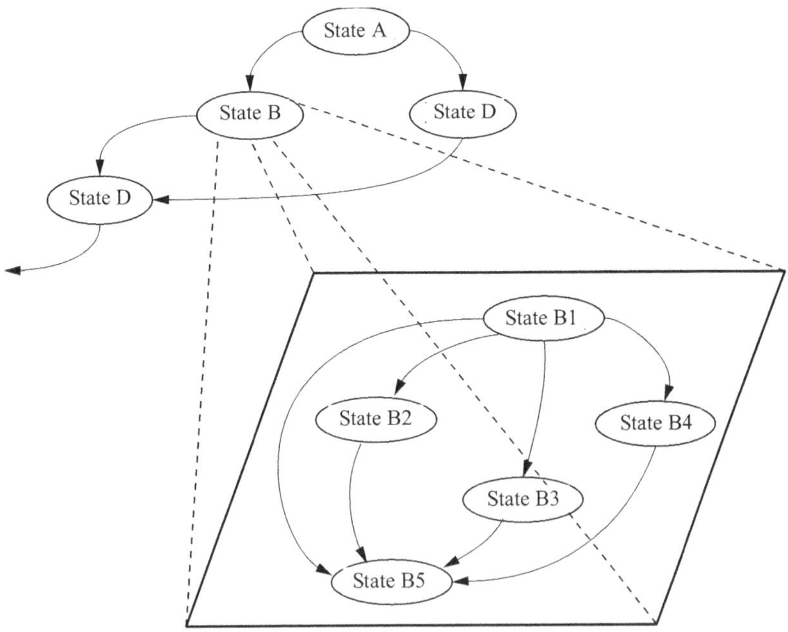

Figure 2.2. *Model arranged hierarchically*

For this, it is possible to create, for a defined level, a state (called a macro-state), which represents a test graph on a lower level of the model (see Figure 2.2). In this way, it is possible to define multiple levels in a model. The inclusion of sub-levels makes it possible to handle a great number of states.

A macro-state is defined as an aggregation of states forming a test graph according to definition 2.2.

Let G1 and G2 be two test graphs and ε ei be a state of the graph G1, the state ei can be refined in a lower level according to the graph G2. Then ei is a macro-state and G2 is regarded as a subgraph of G1. This concept of including the subgraph by a graph leads us to redefine the concept of a test graph. See definition 2.2.

DEFINITION 2.2.– *The test graph* G *is defined by* G = {E, d(e), Start, End, T, A_I, A_O, δ} *such that*:

– E *is a finite number of states*;

– Start ∈ E *is the initial state of the system; this state is not connected to an incoming transition*;

– End ∈ E *is the final state of the system; this state is not connected to an outgoing transition*;

– T = {t_1,... t_M} *is the set of transitions*;

– A_I = {i_1,... i_{Ni}} *is the set of input vectors*;

– A_O = {o_1,... o_{No}} *is the set of output vectors*;

– δ (e, i) *is the transfer function that makes it possible to change from a stable state in the test graph, by following the transition* t ∈ T: E × A_I → (o_1,o_2,..o_n) ∈ A_O × E, *the transition* t ∈ T *is associated with an input vector* i_m ∈ A_I *and a set of output vectors* (o_1,o_2,..o_n) ∈ A_O;

– d(E_S) *is the associating function in a state* e_i ∈ E_s (*subset of the macro-states* E), *a graph* G_i.

A test model can be defined based on the text graph, definition 2.2.

DEFINITION 2.3.– *A test model consists of a set of test graphs linked by hierarchical relations of inclusion.*

The advantages of constructing the test model according to hierarchical levels vary:

– the model is more comprehensible from a graphic point of view;

– when the same sequence of functionalities appears on several occasions, creating only one graph is sufficient and it could be used in various places in the model;

– factoring calculations on the entire model is simplified by the modular approach;

– storing the model in the form of files and re-using them is facilitated.

2.2.2.4. *Inclusion graph*

Using hierarchical levels creates an inclusion graph concerning the model. This inclusion graph defines the dependencies between the model's various hierarchical levels.

DEFINITION 2.4.– *The chain source is the graph corresponding to the root of the inclusion graph. The other constitutive graphs of the test model are indicated under the term subchain.*

To avoid creating infinitely sized models with this inclusion system, we specify the following rule when adding a new subchain in a model: "it is possible to include a completely new chain that does not appear in the source string path, or does not include another subchain, which is already in the source string path".

In other words, it is an acyclic graph. These rules of inclusion are illustrated in the example given in Figure 2.3.

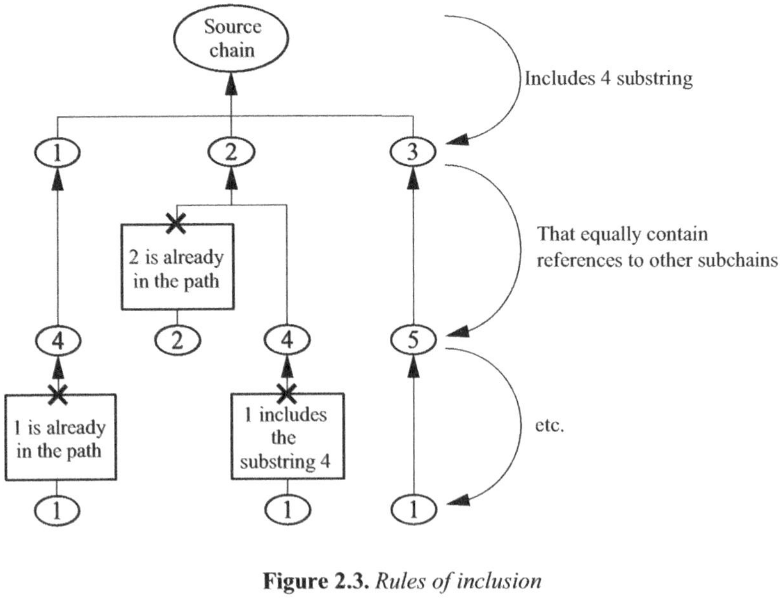

Figure 2.3. *Rules of inclusion*

The relation of inclusion is represented by an arrow going from the subchain to the higher-level string to which it is attached. Figure 2.3 shows subchains that are impossible to include at various places in the inclusion graph, a trigger. The reason for prohibiting inclusion is then represented on the link connecting the subchain to the higher level string.

If the same subchain appears several times in the same model, there will then be two possibilities to consider various calculations (for example, those involving coverage).

It is indeed entirely possible to treat the various graph representations either as only one graph or, on the contrary, as different graphs.

2.2.2.5. *Profile concept*

The reliability is the probability that a specific user will not encounter a failure during system use at a given time. For this reason, it is necessary to be able to handle several profiles of the same system to guarantee the reliability for each type of user.

To build a profile of system use, each input vector is associated with a probability. Thus, each transition is associated with the sum of the probabilities of selection of the input vectors that are associated with it.

With the help of some modifications [LEG 05], each graph included in the model can therefore be compared to a Markov chain in discrete time.

2.2.2.6. *Distributions of the input vectors*

As we saw previously, the input vectors are associated with the model's transitions. Each input vector, moreover, is associated with a probability of being selected. The input vectors associated with the same transition have the following properties:

– they are described according to the same type;

– whatever the input vector chosen for the transition, the state where the chain is found is identical;

– the probability of the transition in the Markov chain is thus equal to the sum of the probabilities of the input vectors associated with this transition.

This modeling involves a probability distribution on two levels: the probability p_{ij} of selecting the transition (i,j) and the probability p_{va} of selecting an input vector knowing that the transition (i,j) was chosen.

PROPERTY 2.1.– The set of input vectors $\{v_1^{(i,j)} \ldots v_A^{(i,j)}\}$ is associated with the transition (i,j). The sum of the conditional probabilities of selection of the input vectors associated with a transition is equal to 1.

$$\sum_{a=1}^{A} P_{V_a}(i,j) = 1$$

As an example, Figure 2.4 illustrates the two considered representations of probability:

– the portion of the figure to the right represents the set of input vectors that can be applied starting from a state of the chain. Each input vector is associated to a transition and the probability of this transition is equal to the probability that a user applies this input vector to the software being tested. This model assumes the creation of six states;

– the portion to the left represents the same example where the input vectors were regrouped during two transitions according to the variable type describing the input vectors.

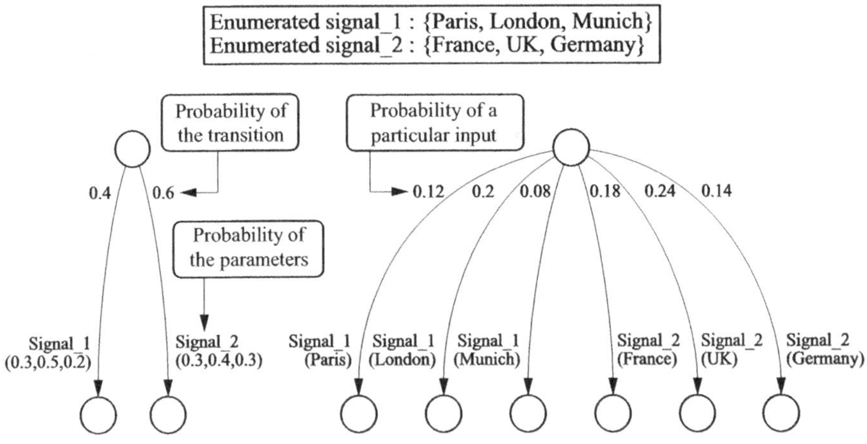

Figure 2.4. *Rules of inclusion*

The factorized representation allows us, on the one hand, to reduce the number of states and transitions of a graph and, on the other hand, to manage quasi-infinite sets of input vectors.

NOTE 2.4.– For the purposes of computing, infinity does not exist. The integer and real variables are transcribed in a finite number of values. However, to be consistent and coherent mathematically, the term quasi-infinite is used to indicate this finite unit that is of too large a size to be enumerated under realistic conditions.

To avoid associating a probability with each input vector, it is necessary to define probability distributions for the various variables associated with a transition.

Variables that are real, integers, a list of elements, and Boolean are considered random variables with well-known probability laws.

For this, consider two types of real random variables:

– rectangular distribution: the probability of selecting an element in a given interval follows a continuous uniform law;

– Gaussian distribution: when a variable is associated with this distribution, it is then necessary to specify its average and its standard deviation.

An integer variable can be associated with a rectangular distribution. Just like a real variable, if the definition set is reduced to only one interval, the distribution is discrete uniform. If the definition set consists of several disjoint intervals, the probability of selecting one interval among others is introduced. After selecting this interval, the conditional distribution inside this unit is then discrete uniform.

For enumerated or Boolean variables, the probability of selecting each item is defined. For "object identifier", "bits chains", "bytes chains", "list of items" or "character string" variables, there is no probability distribution; the values that take these variables are given during the creation of the input vector.

The probabilities of selecting the input vectors are calculated then according to the variable distributions that define them.

2.2.2.7. *Automatic allocation of probabilities*

The allocation of probabilities can be made manually or semi-automatically. It is not limited to just the model transitions, but instead also applies to the set of the data born by a transition, i.e. the input vectors.

It is possible to automatically affect the probabilities according to predefined objectives, such as automatically allocating probabilities to allow an identically intense test for the whole application, or to facilitate and optimize the coverage criteria.

An automatic allocation of probabilities is useful for several levels:

– during the construction of a new model when the usage probabilities are unknown, the automatic allocation of probabilities makes it possible to allot profiles by defect (interesting for the generation) and thus enables system validation;

– with a set of data insufficient for fully determining a profile, automatically allocating probabilities allows the creation of a profile respecting a certain number of known constraints, all the while carefully choosing the missing probabilities. These new profiles created automatically allow for the completion of profile sets already associated with the model for generating new test sequences.

2.2.3. *Principles of generation*

Generation techniques in test cases are diverse:

– the set of test sequences satisfying an objective;

– the set of test sequences offering complete coverage of a reference;

– the set of test sequences that are random.

Once the model is created, it is possible to use it like a support for generating a test case. Several strategies are possible: either deterministic or very random.

The following are some examples of generation algorithms. Several other solutions exist, such as generating the most probable test case or even a test case with input vector values at the bounds of defined intervals.

2.2.3.1. *Random generation*

DEFINITION 2.5.– *Let E be the set of states of a Markov chain,* $E = \{e_1, .., e_Z\}$*. A path C is defined as a finite set of ordered transitions beginning with the transition leaving the* {Start} *state and finishing with the transition arriving in the* {End} *state. The path corresponds to the different transitions encountered during the realization of a process, knowing that the creation of a path stops at the first instance of the* {End} *state.*

A transition is associated with at least one input vector characterized by an instantiation of the various variables associated with the input.

$V = \{x_1,\ldots x_H\}$ indicates the list of variables characterizing the set of input (or output) vectors associated with a transition. An input vector is defined then as $va = \{x_{1,a}., x_{H,a}\}$.

The input vectors associated with the same transition have very similar properties; first, because the variables describing the input vectors are identical (only the data change) and, second, because the application of an input vector does not modify the state (characterized by the set of input vectors that can be applied when the chain is in the state indicated by the transition) of the chain.

Thereafter, the behavior of the system should be (in the absence of error) strongly similar to whatever the vector was that was applied during the transition triggering.

DEFINITION 2.6.– *A test sequence is defined as an ordered set of input and output vectors met during the construction of a path.*

This generation makes it possible to obtain usage sequences close to the behavior of a user insofar as the probabilities of the profile used are representative of reality.

If the fact that the model consists of several graphs is considered, then it is necessary to define how the test sequences will account for the hierarchical dimension.

DEFINITION 2.7.– *A point of origin of generation is one of the {Start} states of one of the constituent graphs of the model.*

In addition, when a macro-state is visited during the creation of the test sequence, the generation is carried out in the corresponding subchain of the {Start} state to the {End} state of the subchain and the generation then resumes in the higher chain exploring the outgoing transitions of the macro-state.

A random generation type can produce identical test cases during a test campaign. Indeed, the interest of the random test is testing according to usage, which simply means that the usage is made up of several identical uses of the software. It is therefore advisable to carry out each test sequence generated randomly.

However, even if the probability of discovering an error is weak while running an identical test sequence, it is not null because initial conditions may vary; this can be compared to the fact that the implementation is not-deterministic. Indeed, the execution of the first test could have modified the internal state of the software, and the execution of the second test will not necessarily provide the same result (for example, a system made up of integrators). Note that many techniques disregard duplicates as they are considered useless.

2.2.3.2. *Coverage of all the states of the chain*

While generating test sequences, it can be interesting to consider the set of test cases of minimal length, which cover all the states in the test model.

This problem is close to the *traveling salesman problem* (TSP), which is stated in the following way:

"A traveling salesman must visit "n" cities, passing through each city only once. He starts with an unspecified city and finishes by returning to the city of departure. The distances between the cities are known. Which course must he choose to minimize the distance covered?"

The problem does not necessarily have a solution. This problem is representative of the class of NP-complete problems. The existence of an algorithm of polynomial complexity remains unknown. The algorithms to solve TSP can be divided into two classes: the deterministic algorithms, which find the optimal solution; and the algorithms of approximation, which provide an almost optimal solution.

Among the existing algorithms, one can quote the *cutting-plane method* [DAN 54], the *facet-finding method* [APP 95], or even the use of a genetic algorithm.

The algorithms of approximation make it possible to find a solution for which the cost is close to the cost of the optimal solution. They have the advantage of making it possible to find a solution within a reasonable time.

2.2.3.3. *Coverage of all the states of the chain*

The problem associated with covering the arcs of a graph is still referred to as the *Chinese Postman Problem (CPP)*. The initial problem consists of identifying the least expensive course through which all the arcs of the graph pass.

A certain number of assumptions can be taken into account; for example, the postman must leave and return at the same determined point, but the number of factors can be higher than one. The graph can be directed if required, and in some cases (for the rural postman) certain arcs are not required to be covered.

Our problem consists of determining a set of test cases, so that each arc appears in this test series. If the graph is Eulerian, i.e. if each state has the same number of arcs entering as leaving, then there is a solution where each arc is covered exactly once. However, this does not have to do with the general case. The problem thus consists of minimizing the additional costs generated by multiple coverage of the same arc.

One solution to this problem consisting of finding a test set that makes it possible to cover all the branches in a directed graph, for a minimal cost, is presented in [LEG 05].

2.2.4. *Some indicators*

Different metrics can be calculated and exploited from the usage model formed by Markov chains.

For example, some basic metrics are:

– number of visits of a state;

– probability of visits of a state;

– probability of return to a state;

– number of returns to a state;

– length of a course.

We can also produce:

– indicators for the test campaign;

– results relating to dependence in a course;

– information on the average number of sequences to visit all the states.

How to calculate all these indicators is detailed in [LEG 05].

2.2.5. *Calculating reliability*

The criteria to stop tests in the industry are seldom based on reliability. However, this quantity is essential to allow calculations of the average time of use before the first failure, and therefore to envisage the cost of non-quality.

Several models of reliability calculation [VAL 04] are limited to the observation of errors and intervals of time between these errors. They are thus based only on the past by making extrapolations on the future. We propose a new solution consisting of the use of information relating to the coverage of functional tests allowing us to guarantee better knowledge of reliability in less time than a test.

An estimate of reliability is calculated for each usage type of the software, i.e. for each profile associated with the Markov chain. These estimates are calculated independent of the profiles used for the generation of the test cases.

We assume that the fact that the reliability is calculated starting from a test model (modeling some particularities of the test architecture) to generate test cases and not a specification model (expressing the functionalities of the application in an operational situation) does not involve a bias when estimating reliability.

Our method of reliability calculation reflects that, in practice, the errors are not immediately systematically corrected. They can be corrected one-by-one or, alternatively, by packages. It also assumes that it is not possible to correct all the errors. This is the case during the delivery of certain software considered to be non-critical or in a trial version.

Our model for the reliability estimation is based on the probability of having an error when triggering a transition because this probability is related to the coverage of the transition. The more important the coverage of a transition, the higher is the probability of not having an error during a new transition triggering. In absolute terms, if the coverage is complete, then the probability of not having an error is equal to one.

The experiment consisting of randomly generating S test sequences is carried out by starting with a given profile that is executed one after the other in the software. When an error is discovered, the execution of the current test is immediately stopped and passed to the following sequence.

During the application of the S tests, corrections of the errors discovered on the software are not made. With each execution of a test sequence, the frequency of activation for each transition and the possible errors discovered are recorded.

This is based on the proposal of K. Sayre [SAY 99], which consists of allotting each transition the probability of having an error at the time of the course of the transition. However, the method of calculating this probability differs from K. Sayre's proposal.

The following stage consists of calculating the probability that an execution is done without error. It is then sufficient to introduce a temporal parameter to obtain reliability.

2.2.5.1. *Probability of having an error during a transition triggering*

We present two solutions for calculating the probability of having an error during a given transition triggering.

The first proposal is based on the number of runs made in this transition, while the second proposal is based on the fact that the classes of equivalence are identified and that the covered classes are stored.

Note that q_i is the probability estimator of having an error while triggering the transition t_i.

Estimator based on the number of runs

This estimator is based on the composition of two indicators to provide the most appropriate result for the probability of having an error during a transition triggering.

The first of the indicators used is based on a Bayesian estimation [SAY 99].

Let $N_i(P_f)$ be the number of runs by the transition ti during the execution of a test series generated randomly starting from a profile P_f. Also n_i* indicates the number of runs by the transition t_i where an error was detected during the execution of this test series, and n_i is the number of runs without error by the transition t_i.

What results is:

$$N_i(P_f) = n_i(P_f) + N_i^*(P_f) \quad [2.1]$$

In this case, the probability $q^1_{i,Pf}$ of having an error during the running of the transition t_i for the profile P_f is calculated using equation [2.2]:

$$q^1_{i,P_f} = \frac{n_i^* + a_i}{n_i^*(P_f) + n_i(P_f) + a_i + b_i} \quad [2.2]$$

The values of the parameters a and b come from data *a priori.*

This result must be calculated for each profile belonging to the set of profiles. Indeed, when the profile for the test generation is changed, it is necessary to know the number of errors generated (which can be very different depending on the profile used for the generation).

The second indicator is based on breaking it down into classes of equivalence. The calculation of the number of classes is based on the number of classes identified thanks to heuristics. This indicator is calculated only when this transition is covered with no error.

C(k) is considered a class of equivalence associated with the transition *ti*, P_f the profile used to generate test sequences, P_{Pf} [C(k)] indicates the probability of selecting the class C(k) according to the profile P_f, $n_i(P_f)$ the number of runs without error by the transition t_i for test cases generated starting from the profile P_f.

The probability r_i that a class C(k) of the transition t_i is not covered at the end of n_i runs by the transition t_i is the product of the probabilities that this class may not be covered for each profile is defined within the framework of equation [2.3]:

$$r_i[C(k)] = \prod_{P_f \in E(P)} (1 - P_{P_f}[C(k)])^{n_i(P_f)}$$

with: [2.3]

$$n_i = \sum_{P_f \in E(P)} n_i(P_f)$$

The hypothesis is that when a class is covered without error, then the application of a new vector of this class cannot cause error.

If it is considered that a class has a probability β of causing an error when it was not covered, then the probability $f_i[C(k)]$ of having an error during the application of a vector of the class C(k) knowing n runs without error is given by equation [2.4]:

$$\begin{aligned} f_i[C(k)] &= \beta.\, r_i[C(k)] + 0.\,(1 - r_i[C(k)]) \\ &= \beta.\prod_{P_f \in E(P)} (1 - P_{P_f}[C(k)])^{n_i(P_f)} \end{aligned} \quad [2.4]$$

NOTE 2.5.– In the absence of information for initialization, it is advised to assign β the value 0.5. This figure can then be updated before each new version according to the proportion of classes not covered, with the preceding version having caused an error while applying a first vector of the considered class.

Then, the probability of having an error during the transition triggering for a user of profile P_f is calculated as the sum of the probabilities of the selection of a class multiplied by the probability that it causes an error as shows in equation [2.5]:

$$\begin{aligned} q^2_{i,P_f} &= \sum_k P_{P_f}[C(k)].\,f_i[C(k)] \\ &= \sum_k P_{P_f}[C(k)].\,\beta.\prod_{P_f \in E(P)} (1 - P_{P_f}[C(k)]^{n_i(P_f)} \end{aligned} \quad [2.5]$$

This result is naturally dependent upon the profile. However, even if no test case was generated according to the considered profile, the result can be calculated for each profile by taking into account the tests carried out; this is not the case for the preceding equation.

Assuming that errors were discovered in this transition, the result could not be expressed because the estimator is based on the expectation of the number of classes covered by considering the number of successful runs. In the absence of errors in the transition, the final estimator is based on the minimum of the two indicators available:

$$q_{i,P_f} = min\,(q^1_{i,P_f}, q^2_{i,P_f}) \quad [2.6]$$

2.2.5.2. *Estimator based on storing the covered classes*

This section presents the probability estimate of not having an error during a transition triggering based on the classes of equivalence and their coverage.

This model is based on the assumption that when a vector of the class is applied without error, then no vector of the class can cause error. The use of this indicator must be limited to the cases where the construction of the equivalence classes is certain.

$f_i[C(k)]$ is known as the probability of having an error during the triggering of the transition t_i, and $P_{Pf}[C\}(k)]$ is the probability of selecting a vector of the class C(k) during the triggering of the transition t_i when the test case is generated starting from the profile P_f. The probability of an error is calculated starting with equation [2.7]:

$$q_i[P_f]=\sum_{k=1}^{k} P_{P_f}[C(k)].f_i[C(k)] \tag{2.7}$$

This value is initialized at the value β at the start of the test campaign. The values $f_i[C(k)]$ will be different in function according to the selected properties to consider an equivalence class as validated. If it is considered that only one validated element of the class leads to the validity of the *simple type coverage class*, then the probability $f_i[C(k)]$ of having an error in the class C(k) takes three distinct values:

– $f_i[C(k)] = b$ if no vector of the class was tested;

– $f_i[C(k)] = 0$ if a vector of the class was tested successfully;

– $f_i[C(k)] = 1$ if a vector of the class was tested and an error was detected.

Remember that errors are not necessarily corrected immediately, which justifies the use of the value one.

When an error is corrected in a class, the probability of discovering an error in this class is assigned the value $\gamma < \beta$, i.e. assuming that the error has a probability equal to γ of being poorly corrected.

We consider that when a class of equivalence is covered without error, the probability that this produces new errors is null; let: $f_i[C(k)] = 0$.

NOTE 2.6.– While the number of errors is important at the beginning of the campaign, if too much information is kept at this stage, the estimate of not having an error will be weakened and it will not progress quickly.

However, if you wish to attain a high reliability, it is necessary to execute a consequent number of tests and to record a sufficiently large history to be able to expect convergence.

This observation leads us to formulate several possible assumptions for choosing a and b:

– it is possible to assume non-regression between two versions of the system, which is often made with the traditional models of reliability estimation. This is to say that the results should be considered only for the last software version;

– you could also decide to grant different weights to information according to their history. On the assumption that quality grows progressively with the test campaign, less weight would be granted to the first software versions where the number of errors was important;

– another solution would be to determine the optimal moment in the history that would enable us to have the maximum probability of not having an error.

We are not giving advice as it relates to choosing a strategy to be applied because the determination of such a choice would require thorough study. However, it is necessary to keep in mind that choosing the parameters is important as regards achieving a fast and correct estimate of reliability. The fact to remember is that covered classes are interested, among other things, in granting a good visibility of the possible profit of reliability in the number of runs made by a transition. It is therefore possible to have good visibility to achieve the stop criterion.

2.2.5.3. *Probability of not having an error during use*

The probability of not having an error is calculated during use for each user profile. It relates to the probability of going from the {Start} state to the {End} state without encountering a defect. To make this calculation, we add an additional transition starting from each state toward a state with an error with a probability equal to the probability of having a defect leaving this state.

Taking into account the fact that a transition is referenced by the two indices i and j, note that q_{ij} is the probability of having an error during the application of an input vector related to the transition t_{ij}.

Note that p^*_{ij} is the new probability of an error-free use with the software in the transition t_{ij}, $p^*_{ij} = (1-q_{ij})p_{ij}$ and the probability of going from state i into the error state is $1 - \sum_j (1-q_{ij})p_{ij}$.

Note that chain F is created after the addition of the error state, and that F' is the matrix obtained starting from F after removing the lines and the columns related to the absorbing states, Pu is the probability that the use of the software is done without error, f is the column of F relating to the {End} state from which the lines related to the absorbing states are removed. The probability of a use with no error is given by equation [2.8]:

$$P_u = (1,0..0)\,(1-F')^{-1} f \qquad [2.8]$$

2.2.5.4. *Calculating the characteristics of reliability*

Once the probability of not encountering an error during the execution of a type of software is known, it is necessary to introduce a temporal variable to obtain reliability.

Let P_u be the probability of not having an error during the execution of the software, μ is the real number expressing the amount of activations of the software per unit of time t. Equation [2.9] presents the calculation of reliability:

$$R(t) = P_u^{ut} \qquad [2.9]$$

Software uses are done independently of each other according to the same profile. Let X be the random variable associated with the behavior of the software during an execution: X = {error, no error}. The probability that there is an error during a software execution is given by equation [2.10]:

$$P\,[X = error] = 1 - P_u \qquad [2.10]$$

Let Y be the random variable recording the number of executions with no error, Y follows an amended geometrical law [TRI 82] and the expectation that Y equals $P_u/(1 - P_u)$. The MTTF is expressed then by equation [2.11]:

$$MTTF = \frac{P_u}{\mu\,(1-P_u)} \qquad [2.11]$$

The MTTF makes it possible to determine the stop criteria of the test campaign adapted to each type of software.

NOTE 2.7.– This method is naturally dependent on the selected profile. To account for the uncertainty that can occur with construction, it is possible to use the work carried out in this field by K. Goseva-Popstojanova and S. Kamavaram using the method of moments [GOS 04].

2.2.5.5. *Interests and limits of this solution*

With this new method, the estimate of reliability is more relevant and realistic. Indeed, reliability is not equal to one, as is the case for the Whittaker model when no error is found, but allows one as the limit when the number of test cases extends towards infinite.

It is not necessary to have good coverage of the chain to have a relevant estimate of reliability. It is enough that the most used functions are sufficiently well tested.

Calculating the probability of error can be carried out irrespective of the test cases executed; they are representative of the use for which it does or does not estimate the probability.

All the test cases can thus be used to modify the probability of having an error in a particular transition. The time necessary to achieve reliability is consequently reduced.

This method accounts for the fact that the probability of discovering errors in functionalities that have been hardly tested or not tested is inevitably stronger than in functionalities that have been greatly tested.

If you consider that a function is represented by a set of transitions in the test model, then the more these transitions are triggered by the tests, the higher is the probability of having an error in these transitions. Thus the probability of having an error in the corresponding function will be weak.

This method is well adapted for calculating reliability because it takes into account the functional structure of the software on the level of its interaction with the user; therefore, thanks to the profile association, it is able to provide a result with a good degree of reliability.

Furthermore, it takes into consideration the importance of an error for a profile because the method does not give it the same weight if it is found during a frequently used transition for this profile.

2.3. The MaTeLo tool

2.3.1. *Engineering tests directed by models, model-based testing*

The software industry has recently lived through a methodological revolution by integrating modeling as an engineering and development tool. The models are used in the phases of design (approach indicated by the term *Model-Based Design*) to allow quasi-systematic methods of generating code starting from models, with tools like Simulink® in the automotive industry and SCADE® in aeronautics or the railway industry.

The test phases, which always represent between 30% and 50% of the development costs, had been ignored quite a while ago by engineering directed by models, until the arrival of the MaTeLo tool developed by ALL4TEC, which, for the past few years, has implemented the *Model-Based Testing* (MBT) approach. To fulfill the reliability standards constraints and the recommendations related to software development defined by standards like CMMI[1] (*Capability Maturity Model Integration*), it is essential to separate the processes of design from the processes of validation and ensure the independence and the impartiality of the test campaigns.

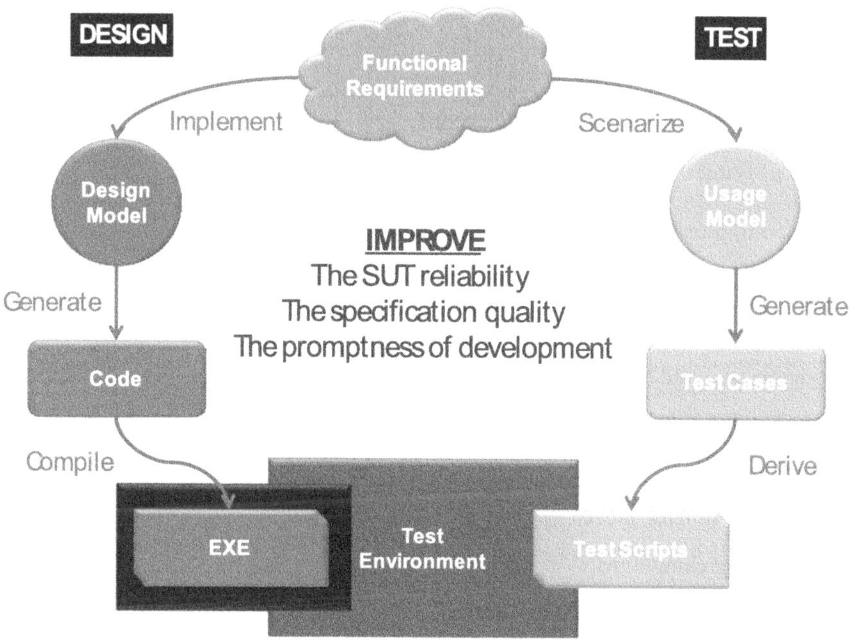

Figure 2.5. *Model-based testing approach in the MaTeLo tool*

For both development and testing, the need is defined by functional requirements. In the MBD approach, the functional requirements are implemented in a design model, aiming to find a solution for the formalized need. In the MBT approach, these functional requirements are expressed in a usage model: the purpose

1 For more information, see www.cmmi.de.

is to validate that the solution sufficiently meets the formalized need. The functional requirements are therefore optimally employed thanks to the use of models, but with different points of view (implementation, writing scenarios) and different objectives (solution, validation) for finalizing the process by evaluating an achievable solution by using test scripts carried out in an automatic test environment.

2.3.2. *A chain of tools*

From modeling to the analysis of the test results, MaTelo offers a complete suite of tools, which consists of the following elements:

– modeling use scenarios (MaTeLo Editor):

- using model design,
- allocating requirements,
- configuring automatic test operations,
- qualifying operational user profiles;

– generating test cases (MaTeLo Testor):

- defining test strategies,
- automatically generating test plans,
- analyzing and reviewing test plans,
- rendering achievable plans in executable test cases;

– carrying out bench tests (externalized via EXAM, NI TestStand, HP QTP, TestNG, SeleniumHQ, etc.):

- planning campaigns,
- automatically executing test cases,
- stimulating and verifying reactions,
- recording results;

– analyzing test results (MaTeLo *Test Campaign Analysis*):

- defining qualitative stop criteria,
- publishing verdicts of executed tests,
- generating reports of analysis for the test campaign,
- analyzing coverage and reliability indicators.

2.3.3. *The usage model*

In MaTeLo, the usage model is the equivalent of a formal specification of the test system requirements. It relates to a model representing all possible stimulations of input of the system interacting in various usage contexts. The model has a statistical component, which corresponds to the input distributions of the system following various laws of possible use.

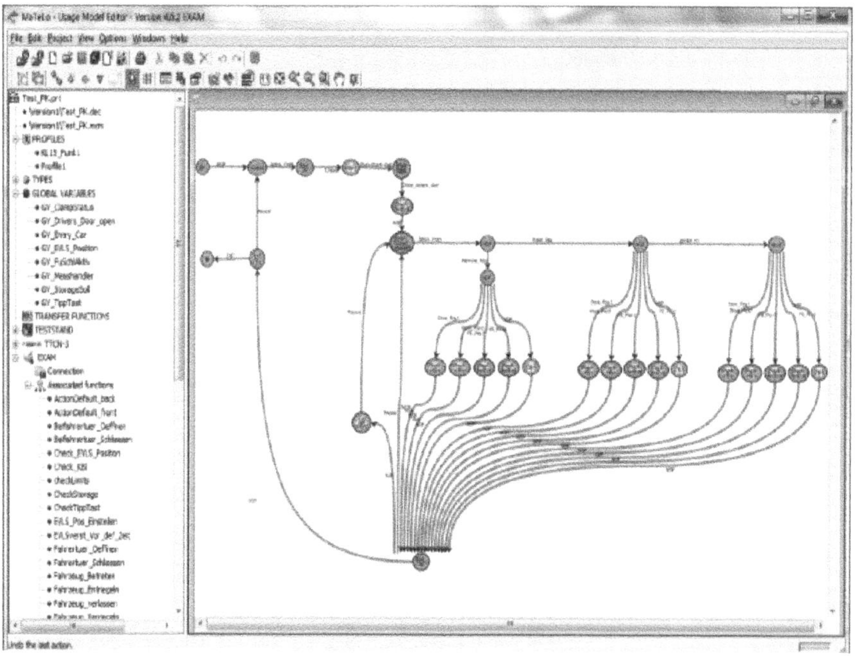

Figure 2.6. *MaTeLo's usage model*

The usage model also makes it possible to ensure the traceability between the functional requirements and the system use scenarios. MaTeLo can check that all the requirements considered functional are implemented by at least one scenario and are testable.

Thus, once checked by a review, such as by peers, the MaTeLo model can be considered a formal specification of the test of the functional system requirements.

2.3.3.1. *Hierarchy of the MaTeLo model*

The MaTeLo tool proposes to treat usage models hierarchically such as that described in section 2.2.2.3.

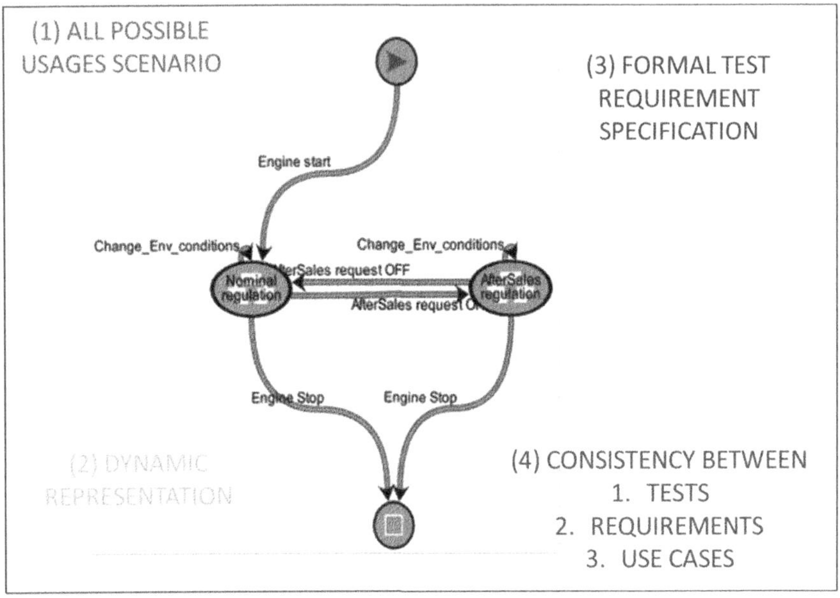

Figure 2.7. *States and transitions in the MaTeLo model*

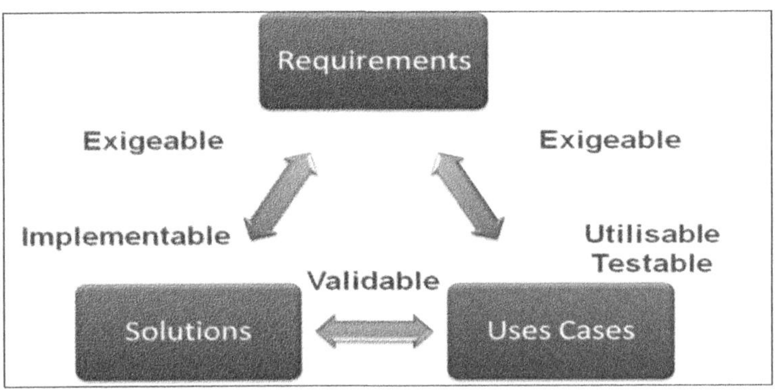

Figure 2.8. *Traceability links of traceability*

Organizing it into a hierarchy offers powerful re-use functionalities of standardized subchains, which are called "macro-chains" in the tool.

2.3.3.2. *The transition as an "atomic test step"*

A transition can be seen as a configurable atomic test step. We associate it with the profiles of its input and the anticipated results of its output.

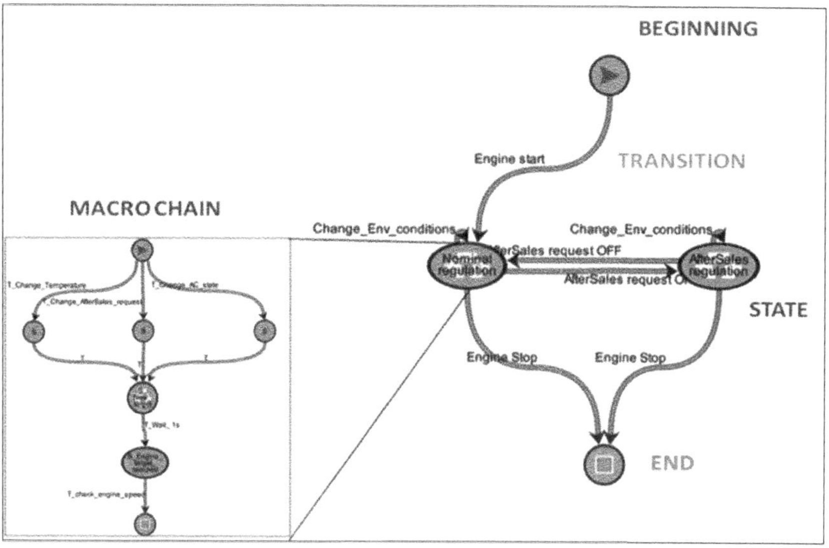

Figure 2.9. *MaTeLo model arranged hierarchically*

The input fields can be partitioned into classes of equivalences, and the models of achievable requirements can be connected to the transitions to calculate the anticipated results automatically and therefore to serve as an oracle.

Once the data of the functional test is modeled, it is necessary to ensure coverage of the requirements (acquired from the basis of DOORS or another tool).

It is necessary to map each variable defined as an input or output with the operational test sequence, which will be then be implemented by a tool for automating tests.

The model makes it possible to define types of variables specific to the project (enumerated, sequences, numerical, character strings) for the instantiation of the input and output, but also for the creation of aggregate variables able to be used to store the history and the context during test generation.

2.3.3.3. *Model input and data classes of equivalence*

The stimulation phase of the system being tested is formalized by several parameters: input, their classes of equivalences, as well as the laws of distribution of the input allocated to each profile:

– the input is the port of entry of the system being tested, which is to be stimulated in the transition to validate the requirements being covered;

– the classes of equivalence are a set of values of the output value field that respect a unique relationship with the output;

– the laws of distribution represent the probability of the appearance of the input data compared to a particular use or "profile".

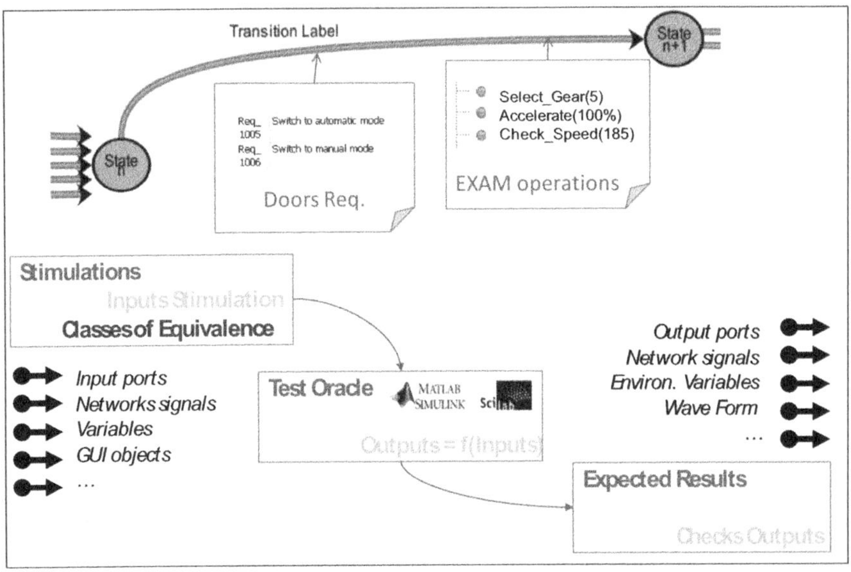

Figure 2.10. *The semantics of MaTeLo transitions*

2.3.3.4. *The oracle test and anticipated results*

The oracle test is a mechanism that makes it possible to automatically calculate the anticipated results when leaving the system. There are several types of oracle tests:

– *Oracle manual*: the originator of the tests formalizes the anticipated results at the time of modeling. For example: Speed_Motor E [3,500 rpm; 3,700 rpm];

– *Oracle memory*: the value of the anticipated result is the result of a system memory updated in a preceding transition using an aggregate variable. For example: Speed_Motor >= VG_MinimunEngineSpeed;

– *Oracle transfer function*: the anticipated result is modeled formally at a very high level (level requirement), dependent on the set of input data, using abacuses (tables or curves of equivalence), or via the simple behavioral models.

Figure 2.13 presents an abacus example, which can be used as an oracle. The curve of equivalence represents the expected results for one or more requirements.

It is possible to very easily model these requirements by using a Simulink© model (logic type or *lookup table*).

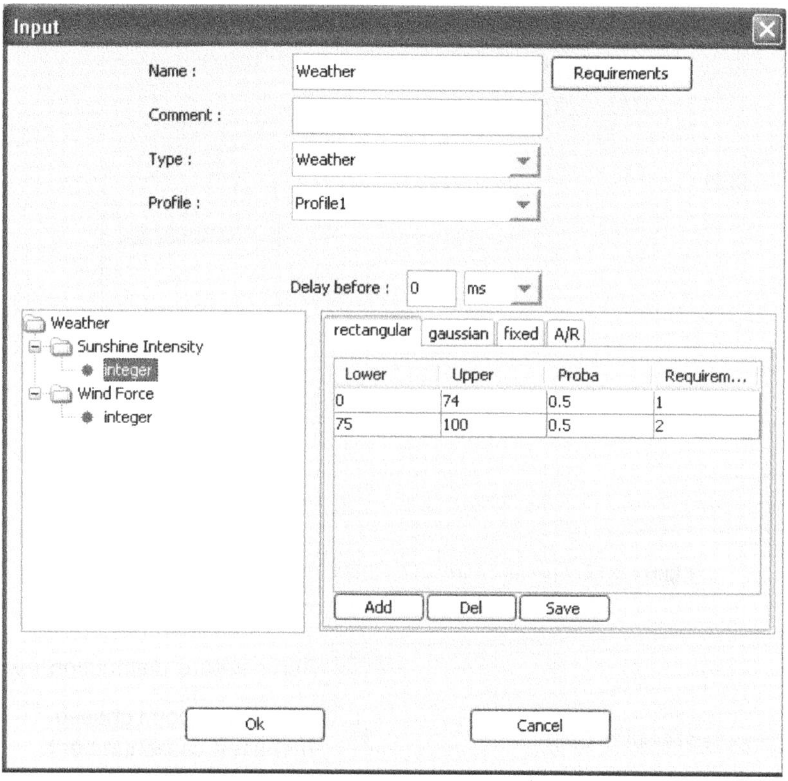

Figure 2.11. *Creation of an input with two classes of equivalence*

MaTeLo allows partitioning the input into five classes of equivalences connected with the Simulink® model. The anticipated results are explicitly calculated according to the input data generated, and the tool for test campaign analysis ensures passage in all the classes of definite equivalences.

The MaTeLo tool allows us to implement, as needed, these different oracle mechanisms.

The tool's performance lies in the fact that oracles are defined by each transition and not on the whole. Oracles related to the stimuli defined by classes of equivalence make it possible to model the totality of the functional aspects of one or more requirements.

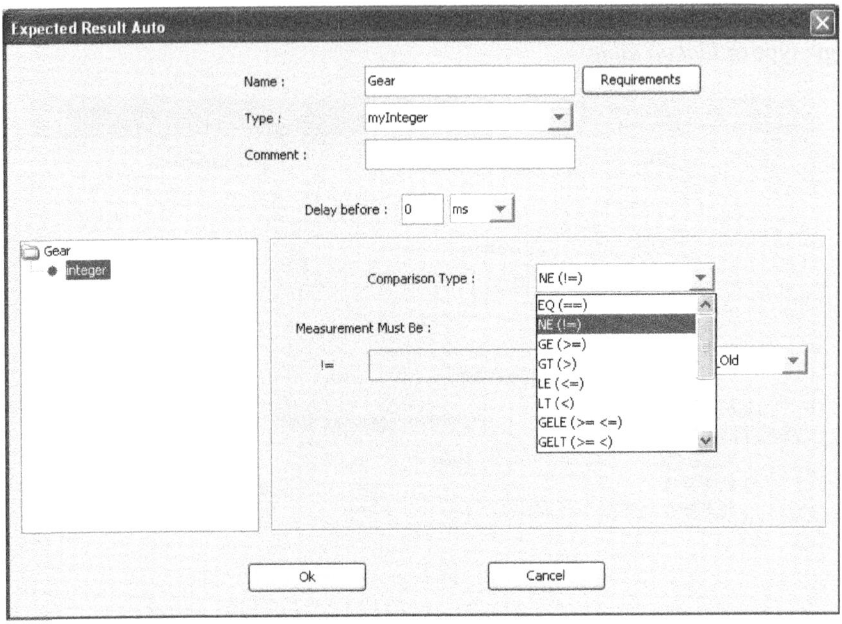

Figure 2.12. *Formalization of the anticipated results in MaTeLo*

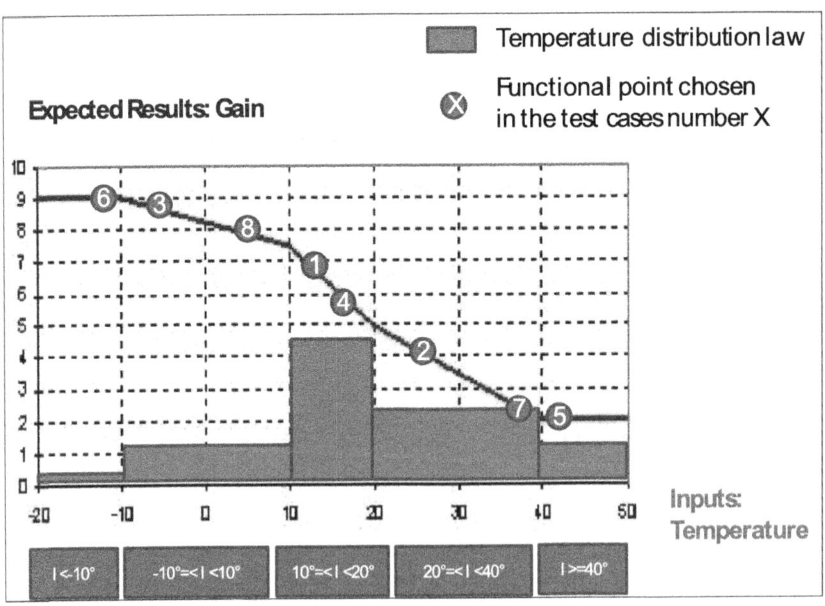

Figure 2.13. *Example of an abacus that can serve as an oracle*

These oracles can be modeled either via arithmetic functions in MaTeLo or via models produced from Simulink® or Scilab tools.

It also provides the possibility of inserting *python* scripts.

2.3.3.5. *Test profiles*

A test profile can be interpreted as a "qualifier" of a usage model. Indeed, it is used to weigh the various possibilities offered by the model, whether it is in terms of path or data input. A weight is defined via the parameters of the laws of distribution to be used.

There are two types of profiles:

– *operational profile*: represents the future uses of the system. It is used to validate operational reliability, among other things;

– *test profile*: represents the acquired priorities of the reflections of the functional or system test analysts.

The profile can be seen as an overlay that is superimposed on the model and that emphasizes more precisely what we want to test.

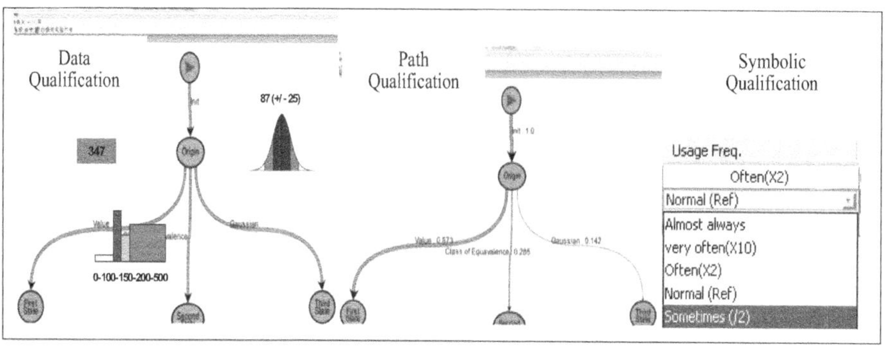

Figure 2.14. *Different test profiles*

2.3.3.6. *Requirements*

The requirements defining the need for validation can be directly:

– defined in the MateLo tool;

– imported from third-party solutions like Rational DOORS or another tool;

– imported via CSV or XML formats.

The requirements are defined by a unique name and associated with a description. The safety requirements can thus be *tagged* to be identified differently thereafter.

The MaTeLo tool makes it possible to trace various associations between the model and the requirements and therefore to ensure that the model covers the requirements.

A requirement can be associated, in a necessary and/or sufficient manner, with the various objects of the model, which are the transitions, input, output, classes of equivalence, or anticipated results.

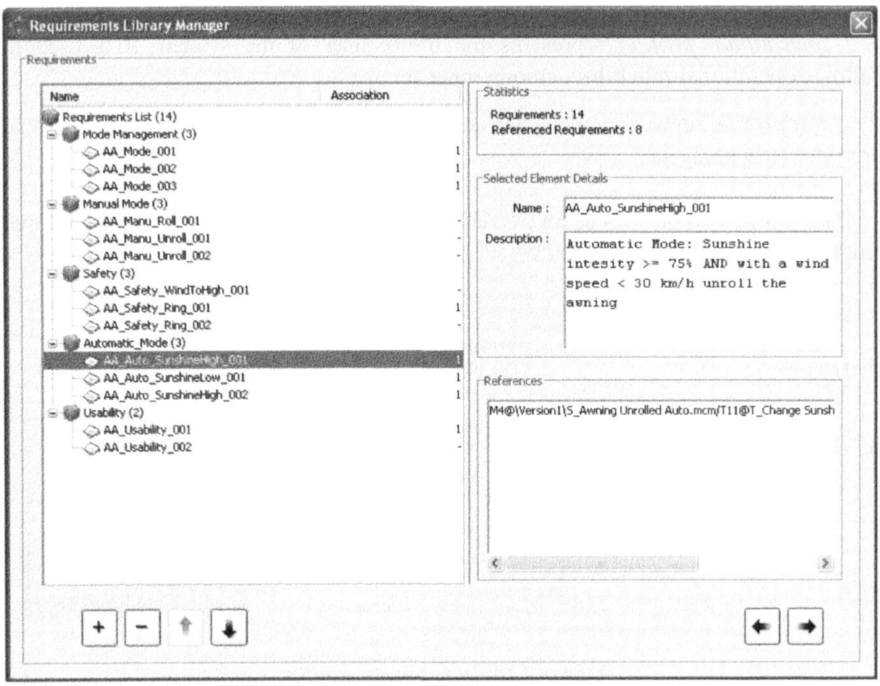

Figure 2.15. *Managing requirements in MaTeLo*

2.3.3.7. *Attributes*

The MaTeLo model allows us to configure specific attributes defined by the project.

These attributes can be associated with the different objects of the model that are the states, transitions, input, classes of equivalence, output, or anticipated results.

With these attributes, each datum of the model can be qualified via additional structured data, thus increasing the quantity of formalized and processed information.

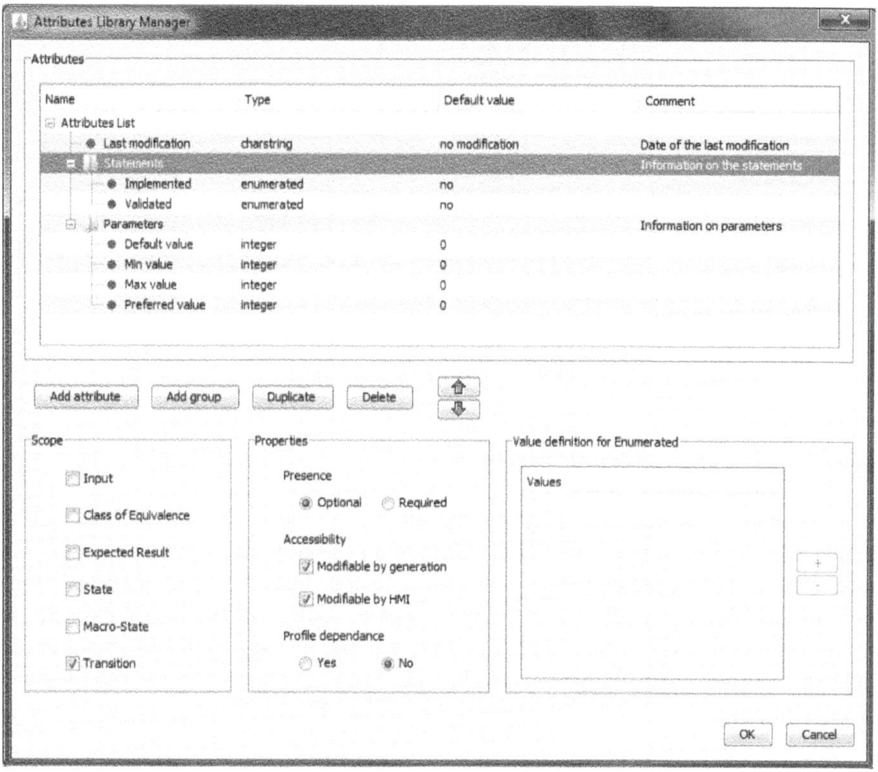

Figure 2.16. *Example of attributes definition*

2.3.3.8. *The test bench*

To allow the MaTeLo tool to generate manageable test cases, it is necessary to map the usage model with its data in the automatic target test environment.

MaTeLo thus makes it possible to take into account the parameters of the automatic test bench and its various primitives, allowing direct access to stimulating input and verifying output or controlling a simulator, measuring apparatus, or qualification.

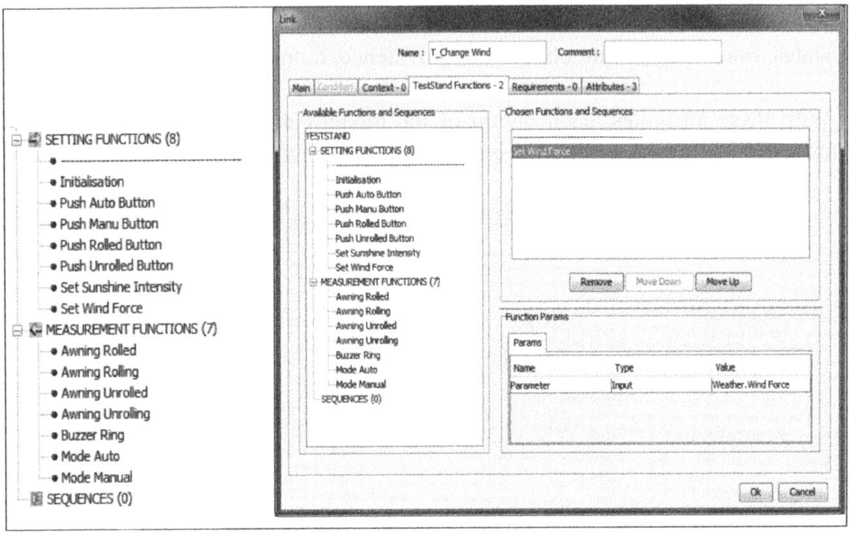

Figure 2.17. *Defining the test bench and mapping the test operations*

2.3.4. *Configuration of test strategies*

Choosing a strategy is very important for properly conducting validation campaigns. In fact, test optimization depends on the test case optimizations to run according to the maturity of the system being tested. Indeed, the nominal test cases have almost no effect on a very mature system, and conversely in an immature system, the test usage cases will highlight too many bugs and will explode the analysis of results. The objective is to find the system's bugs at a useful moment and in a useful order.

The strategy described within the box in Figure 2.18 is that most often implemented.

The strategy presented in Figure 2.17 tests the nominal functions of the system according to several operational profiles, and then makes trade tests based on the complexity or criticality of updates until arriving at optimal coverage tests. Then, when everything is covered at least once in a context, it is interesting to generate test cases whose objective it is to cover the system usage and to validate its reliability.

These test strategies are configurable and can evolve with time via new algorithms. They are defined by:

– choosing an algorithm;

– selecting a profile;

– selecting a part or the totality of the usage model that is to be taken into account.

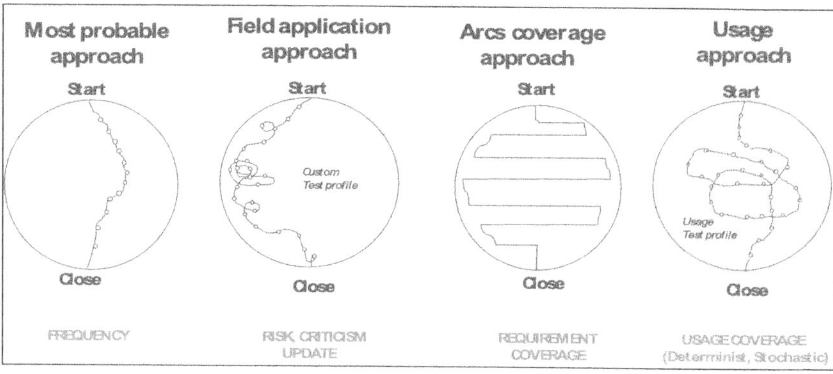

Figure 2.18. *"Usual" sequence of test strategies*

2.3.5. *Generating test campaigns*

By respecting the test strategy used, MaTeLo generates test cases and posts a synthetic report of generation, which makes it possible to select the most interesting test cases by evaluating their coverage or by carrying out reviews of scenarios.

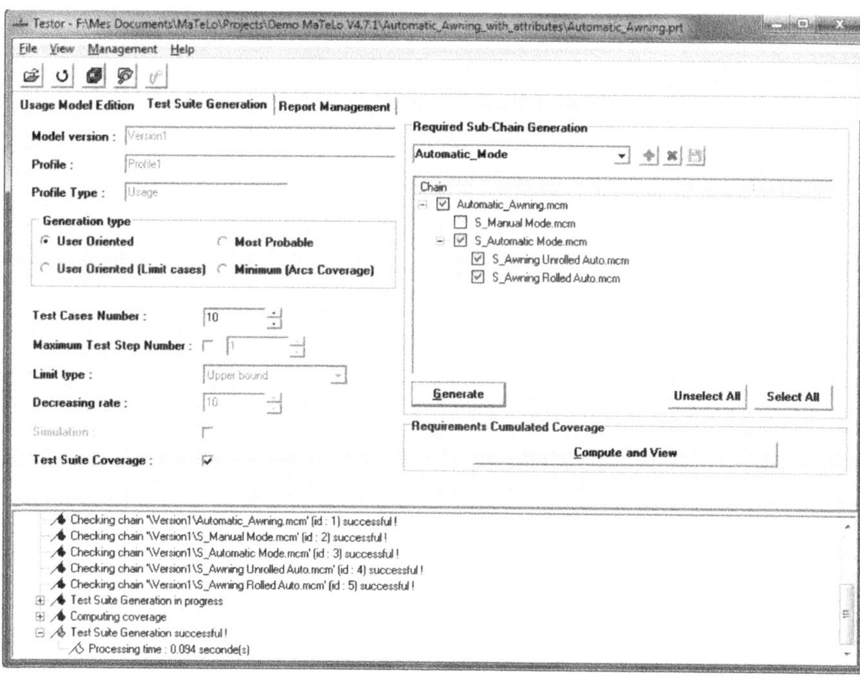

Figure 2.19. *Definition of a test strategy*

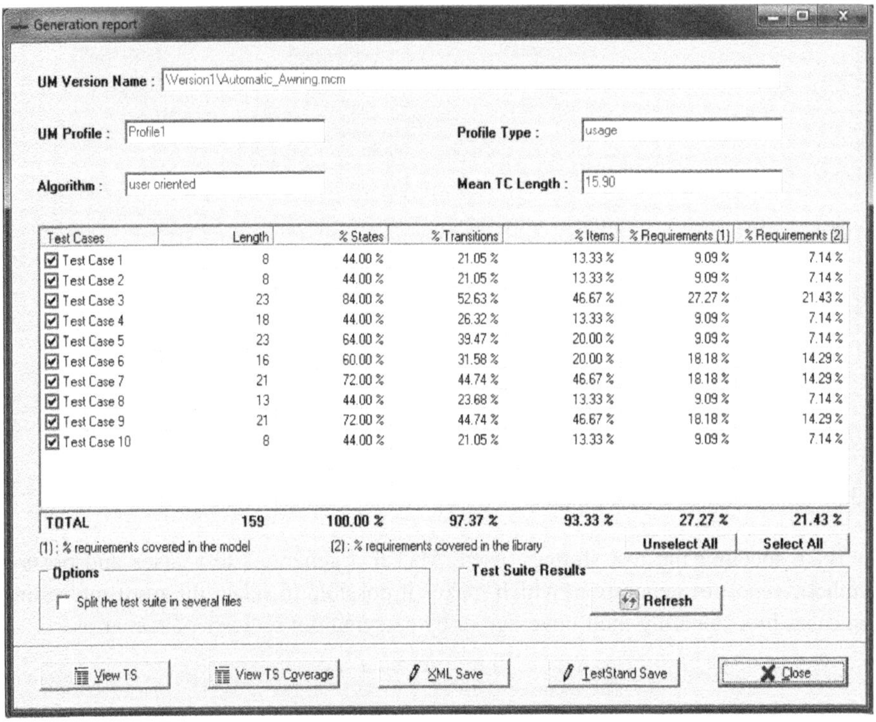

Figure 2.20. *Report of test generation*

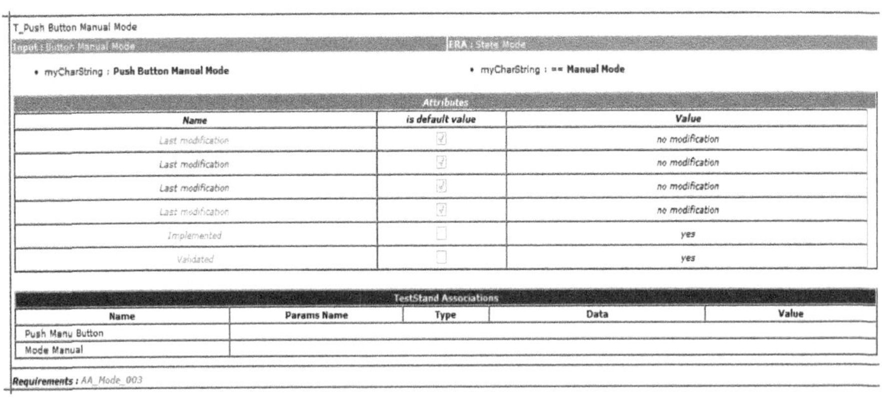

Figure 2.21. *Example of a stage of a test file "in manual mode"*

Reviewing test cases can be carried out by posting the test files "in manual mode", which is presented within the box in Figure 2.21. There we can find:

– the input and their associated data;

– the output and their anticipated results;

– information contained in the attributes;

– configuration of the test operations (here TestStand);

– and coverage of the requirements.

For more details on the coverage of test cases compared to the requirements, the tester can require the MaTeLo tool to post the report presented within the box in Figure 2.22.

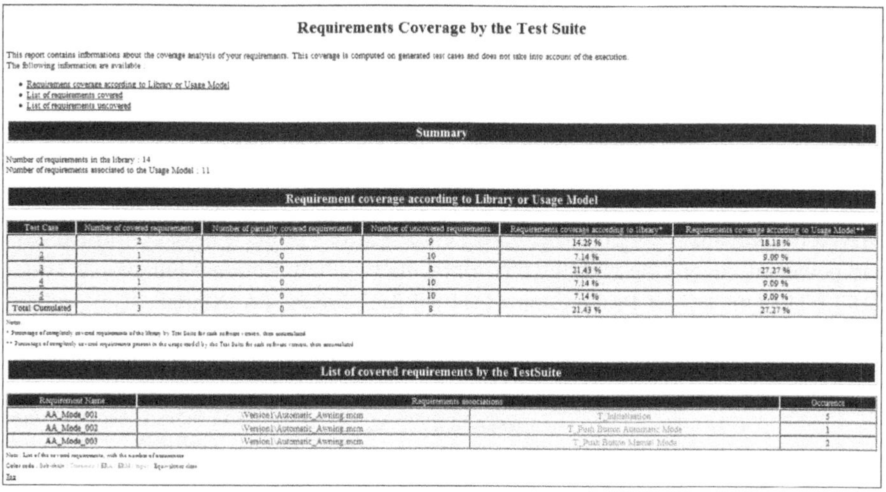

Requirements Coverage by the Test Suite

This report contains informations about the coverage analysis of your requirements. This coverage is computed on generated test cases and does not take into account of the execution.
The following information are available :

- Requirement coverage according to Library or Usage Model
- List of requirements covered
- List of requirements uncovered

Summary

Number of requirements in the library : 14
Number of requirements associated to the Usage Model : 11

Requirement coverage according to Library or Usage Model

Test Case	Number of covered requirements	Number of partially covered requirements	Number of uncovered requirements	Requirements coverage according to library*	Requirements coverage according to Usage Model**
1	2	0	9	14.29 %	18.18 %
2	1	0	10	7.14 %	9.09 %
3	3	0	8	21.43 %	27.27 %
4	1	0	10	7.14 %	9.09 %
5	1	0	10	7.14 %	9.09 %
Total Cumulated	3	0	8	21.43 %	27.27 %

List of covered requirements by the TestSuite

Requirement Name	Requirements associations		Occurence
AA_Mode_001	\Version1\Automatic_Awning.mcm	T_Initialisation	5
AA_Mode_002	\Version1\Automatic_Awning.mcm	T_Push Button Automatic Mode	1
AA_Mode_003	\Version1\Automatic_Awning.mcm	T_Push Button Manual Mode	2

Figure 2.22. *Detailed report of test coverage*

2.3.5.1. *Supported bench tests*

MaTeLo is used today in various fields of industry. It interfaces particularly well with the automatic bench tests used in the following sectors:

– industry (Figure 2.23);

– transport (Figure 2.23);

– telecommunication and information systems (Figure 2.24).

MaTeLo allows the automatic generation of test cases for the following environments:

– manual test: configurable test sheets in XML/HTML formats;

– EXAM: tool for automation published by MicroNova and mainly used by Volkswagen. This tool is employed by more than 500 users. It is free and makes it possible to describe the test cases without giving their implementation details. All the data is stored in bases and is accessible collaboratively (Figure 2.25). In MaTeLo Editor, the test originator can directly map test operations available in the base with the model's data, and thus can automatically generate operational test cases in this base;

– NI TestStand: test sequencer developed by NI and used by many companies around the world in various branches of industry. This tool is employed by more than 1,000 users. It allows the tester to sequence and configure the test steps coded in LabWindows/CVI, LabView, or third-party environments. MaTeLo makes it possible to define the test steps, their mapping with the data, and takes over the TestStand application to create test cases in accordance with the generation;

– proprietary benches: the other benches can be easily interfaced with MaTeLo thanks to an easy integration carried out by the user. The MaTeLo tool is capable of generating all the types of texts and known codes (C, c#, java, vb, python, ruby, perl, etc.) or all types of proprietary languages that do or do not support the Keyword Driven formalisms (CSV, HTML, XML, etc.). Various projects have confirmed easy and powerful integration with the environments EggPlan, Rational Functional Tester (RFT), HP Quick Test Professional (QTP), PROVEtech: TA, Matlab, jUnit, nUnit, TestNg, SeleniumHQ, etc.

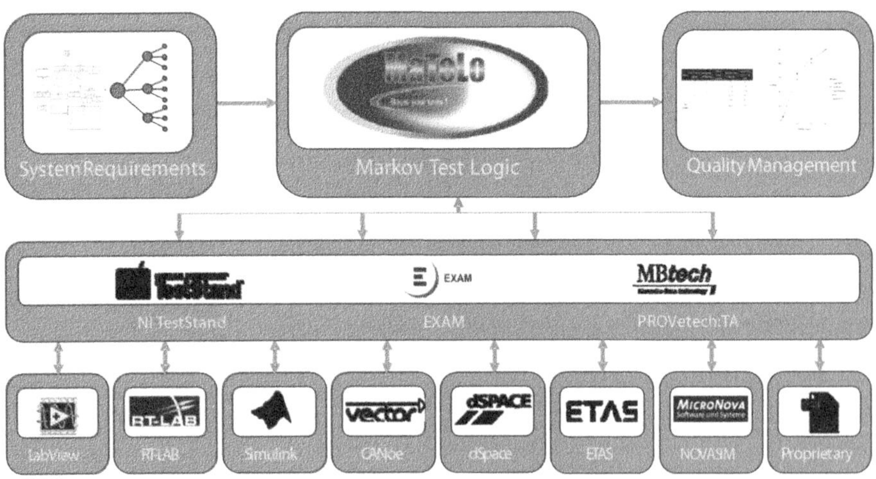

Figure 2.23. *MaTeLo and the benches used in industry and transport*

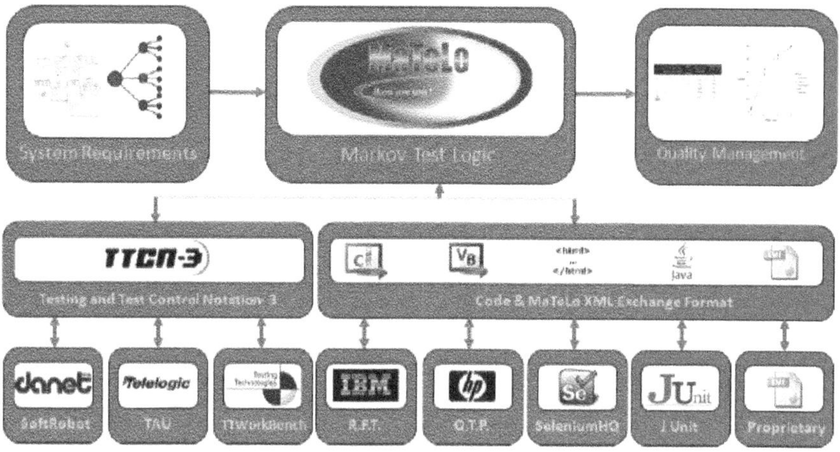

Figure 2.24. *MaTeLo and the environments used for telecommunication and information systems*

2.3.5.2. *Executing tests and verdicts*

The execution of test cases is delegated to the test environment. The only return necessary and sufficient for MaTeLo by test case is as follows:

– status: Not Run/Passed/Failed;

– if there is an error: Error step (stage of the test case or the first failure has appeared).

2.3.6. *Analysis of the results and indicators*

The analysis of the test campaign fulfills the quality assurance of the system being tested by posting, in particular, the test coverage carried out according to the system versions being tested. There are three levels of reports (expert, developer, manager) posting information, more or less.

The first thing to be defined by the user is the set of stop criteria, which will be used to evaluate the difference between the achieved quality and the objective.

The following objectives are proposed in the MaTeLo tool:

– objectives of coverage of the requirements, states, transitions, classes of equivalence, and totality: in a unitary (through a version of SUT) or cumulated (throughout all the versions of SUT) manner;

– objectives of reliability: probability of failure, *Mean Time To Failure*, and failure intensity by operational profile defined in the model.

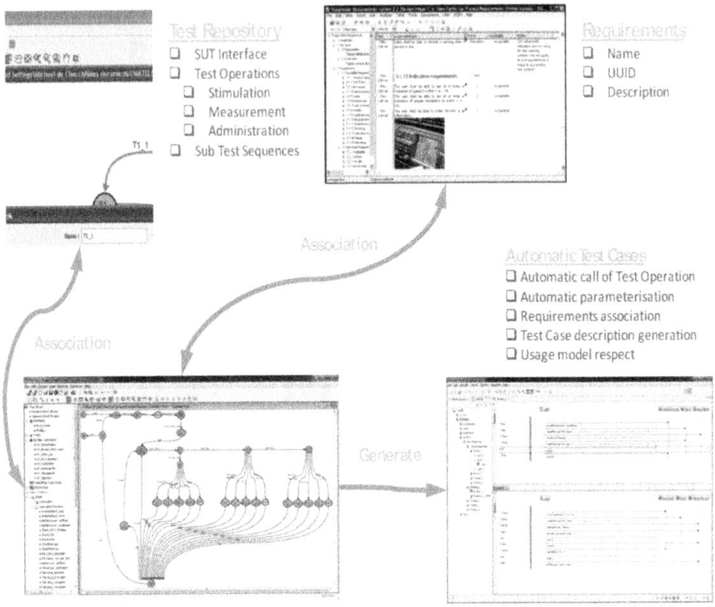

Figure 2.25. *MaTeLo/EXAM connection*

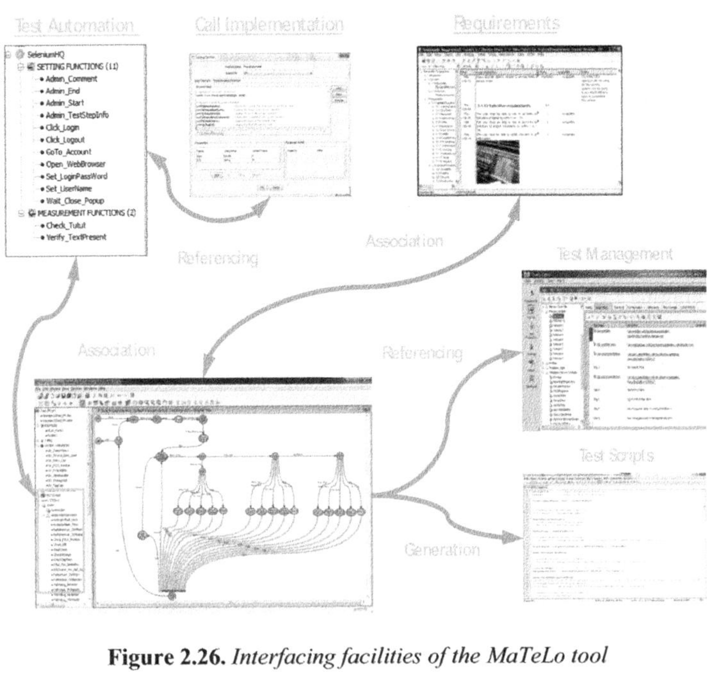

Figure 2.26. *Interfacing facilities of the MaTeLo tool*

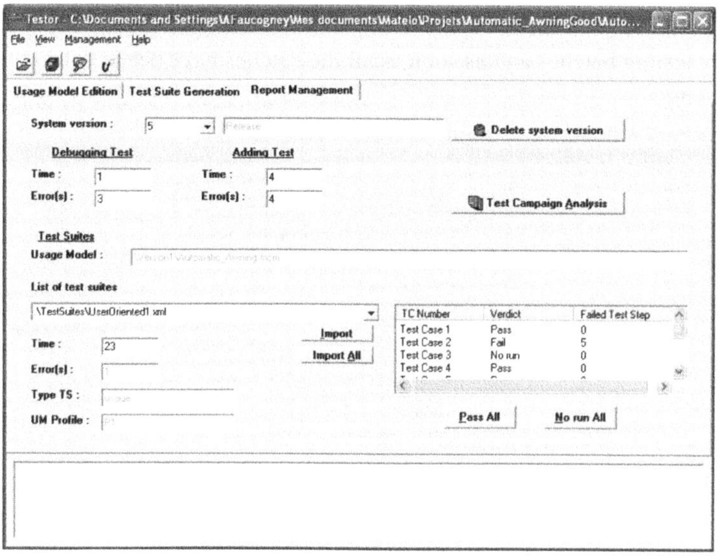

Figure 2.27. *Results of running a test campaign*

At the end of each test campaign, and starting from the model, test verdicts, and stop criteria, the MaTeLo test campaign analyzer proposes to the user various indicators presented in reports. This information is delivered in two ways:

– unitary: the indicators provide the effective coverage for a given version of the system being tested;

– cumulated: the indicators are cumulated throughout all the versions of the system being tested.

Figure 2.28 presents the various indicators available.

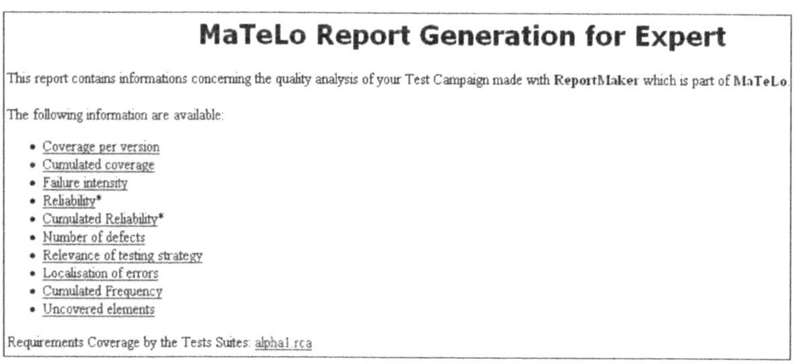

MaTeLo Report Generation for Expert

This report contains informations concerning the quality analysis of your Test Campaign made with **ReportMaker** which is part of **MaTeLo**

The following information are available:

- Coverage per version
- Cumulated coverage
- Failure intensity
- Reliability*
- Cumulated Reliability*
- Number of defects
- Relevance of testing strategy
- Localisation of errors
- Cumulated Frequency
- Uncovered elements

Requirements Coverage by the Tests Suites: alpha1.rca

Figure 2.28. *List of various indicators proposed by MaTeLo*

Thanks to these indicators, the user can, for example, adopt the strategy of separately testing batches and as soon as all the batches have been covered, retest the entire system.

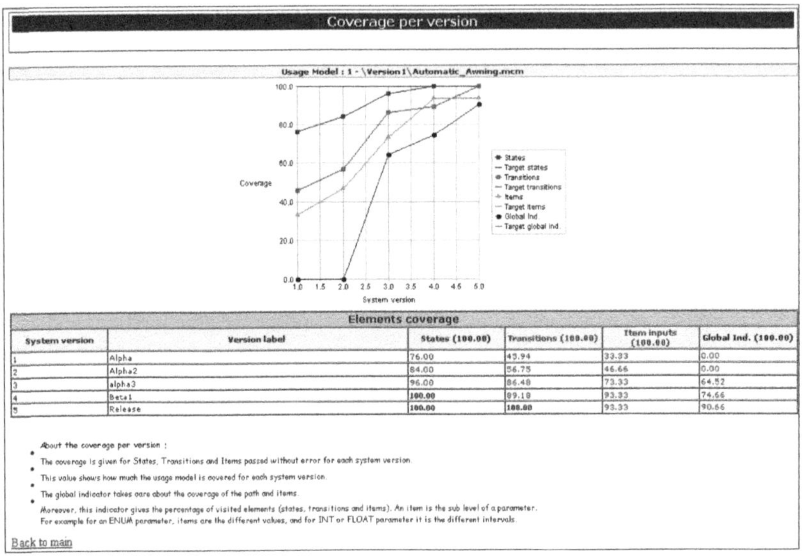

Elements coverage					
System version	**Version label**	**States (100.00)**	**Transitions (100.00)**	**Item inputs (100.00)**	**Global Ind. (100.00)**
1	Alpha	76.00	43.94	33.33	0.00
2	Alpha2	84.00	56.75	46.66	0.00
3	alpha3	96.00	86.48	73.33	64.52
4	Beta1	**100.00**	89.18	93.33	74.66
5	Release	**100.00**	**100.00**	93.33	90.66

About the coverage per version :

- The coverage is given for States, Transitions and Items passed without error for each system version.
- This value shows how much the usage model is covered for each system version.
- The global indicator takes care about the coverage of the path and items.
- Moreover, this indicator gives the percentage of visited elements (states, transitions and items). An item is the sub level of a parameter. For example for an ENUM parameter, items are the different values, and for INT or FLOAT parameter it is the different intervals.

Back to main

Figure 2.29. *Indicator of coverage by version*

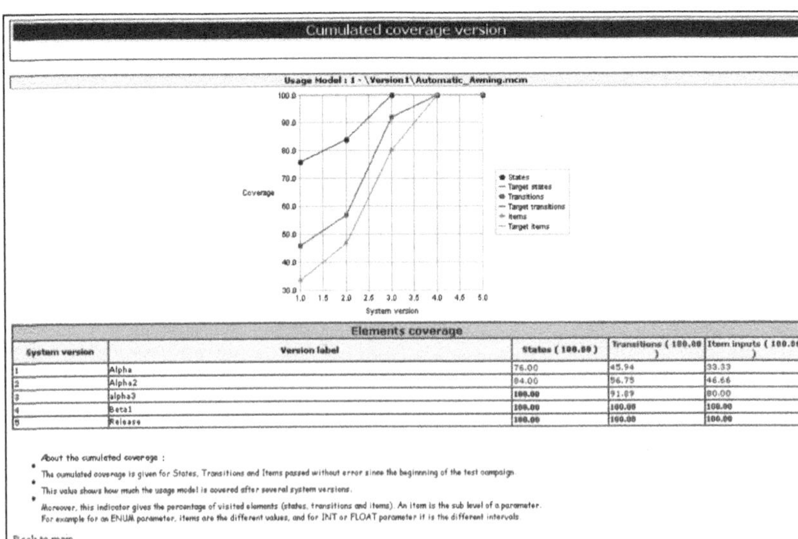

Elements coverage				
System version	**Version label**	**States (100.00)**	**Transitions (100.00)**	**Item inputs (100.00)**
1	Alpha	76.00	45.94	33.33
2	Alpha2	84.00	56.75	46.66
3	alpha3	**100.00**	91.89	80.00
4	Beta1	**100.00**	**100.00**	**100.00**
5	Release	**100.00**	**100.00**	**100.00**

About the cumulated coverage :

- The cumulated coverage is given for States, Transitions and Items passed without error since the beginning of the test campaign.
- This value shows how much the usage model is covered after several system versions.
- Moreover, this indicator gives the percentage of visited elements (states, transitions and items). An item is the sub level of a parameter. For example for an ENUM parameter, items are the different values, and for INT or FLOAT parameter it is the different intervals.

Back to main

Figure 2.30. *Indicator of cumulated coverage*

The delivered indicators can be classified into various categories:

– *coverage of the functions and uses* (Figures 2.29 and 2.30):

- coverage of the states, transitions, and classes of equivalence. Global indicator (taking into account the coverage history, the classes of equivalence, the preceding paths and their coverage progression speed);

– *coverage of the requirements*:

- coverage of the requirements by a version of SUT (Figure 2.31),

- coverage of the requirements by test campaign (Figure 2.32);

– *indicating confidence*:

- reliability and *Mean Time To Failure* are generated as shown in Figure 2.33;

– *other indicators*:

- placing errors in the model (Figure 2.35),

- productivity of the test strategies (Figure 2.34),

- number of runs per object of the model.

Requirements Coverage Analysis by the Executed Tests

This report contains informations about the coverage analysis of your requirements.
The following information are available :

- Requirement coverage according to Library or Usage Model
- List of covered requirements
- Requirements traceability

Summary

Usage Model Version: 1
Number of requirements in the library : 8
Number of requirements associated to the Usage Model : 8

Requirement coverage according to Library or Usage Model

Usage Model	System Version	Label Version	Number of covered requirements	Number of partially covered requirements	Number of uncovered requirements	Requirements coverage according to library*	Requirements coverage according to Usage Model**
Version1	1	Alpha	0	0	8	0 %	0 %
	2	Alpha2	0	0	8	0 %	0 %
	3	alpha3	4	2	2	50 %	50 %
	4	Beta1	8	0	0	100 %	100 %
	5	Release	8	0	0	100 %	100 %
	Total Cumulated		8	0	0	100 %	100 %

Notes:
* Percentage of completely covered requirements of the library by the Test Case
** Percentage of completely covered requirements present in the usage model by the Test Case

Figure 2.31. *Indicator of total coverage of the requirements*

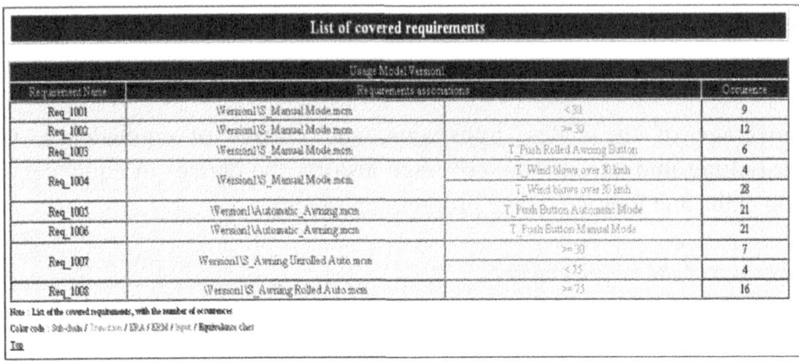

Figure 2.32. *Indicator of coverage of the requirements by campaign*

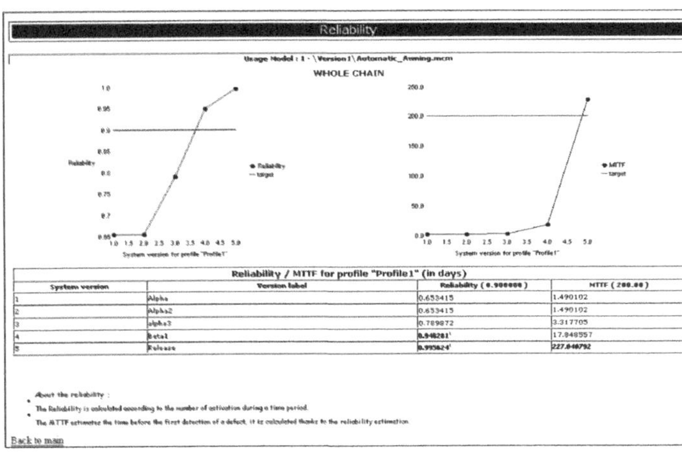

Figure 2.33. *Indicators of reliability*

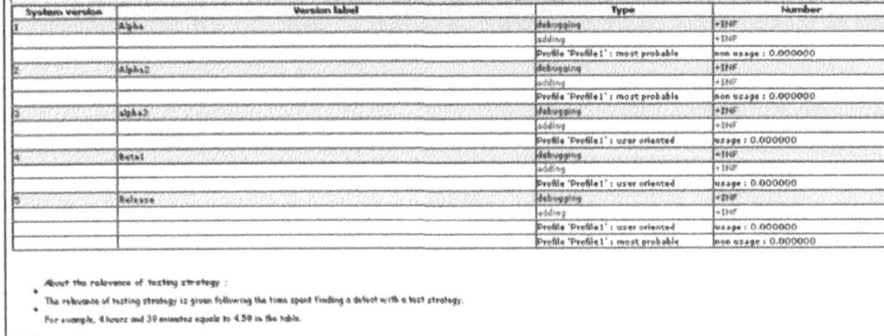

Figure 2.34. *Productivity of the test strategies*

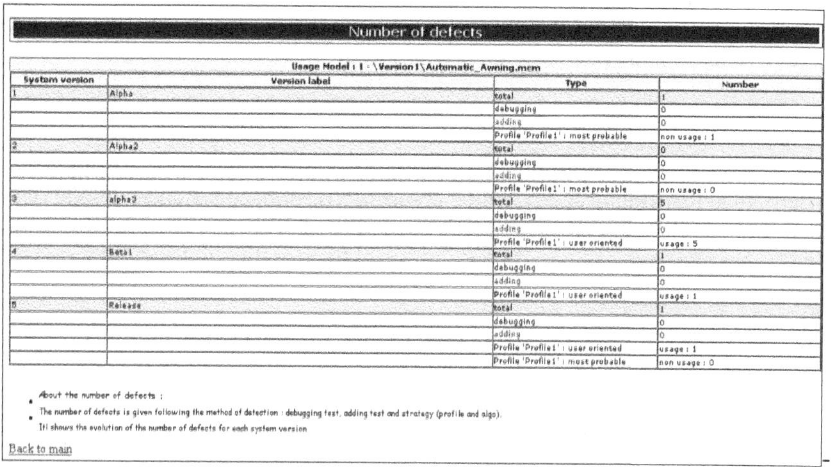

Number of defects

Usage Model : 1 - \Version1\Automatic_Awning.mcm

System version	Version label	Type	Number
1	Alpha	total	1
		debugging	0
		adding	0
		Profile 'Profile1' : most probable	non usage : 1
2	Alpha2	total	0
		debugging	0
		adding	0
		Profile 'Profile1' : most probable	non usage : 0
3	alpha3	total	5
		debugging	0
		adding	0
		Profile 'Profile1' : user oriented	usage : 5
4	Beta1	total	1
		debugging	0
		adding	0
		Profile 'Profile1' : user oriented	usage : 1
5	Release	total	1
		debugging	0
		adding	0
		Profile 'Profile1' : user oriented	usage : 1
		Profile 'Profile1' : most probable	non usage : 0

- About the number of defects :
- The number of defects is given following the method of detection : debugging test, adding test and strategy (profile and algo).
  Itl shows the evolution of the number of defects for each system version

Back to main

Figure 2.35. *Placing errors in the model*

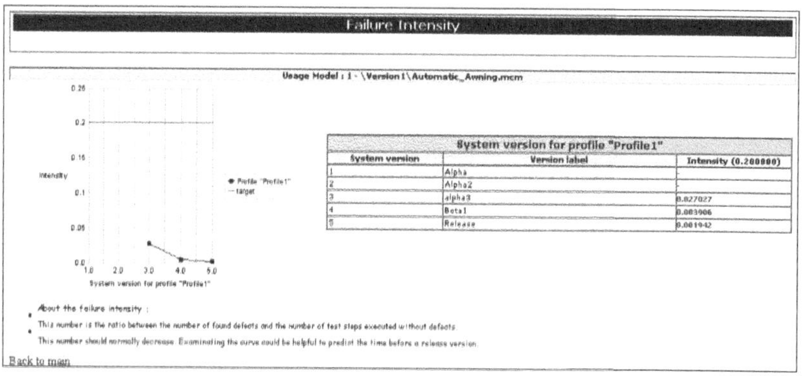

System version for profile "Profile1"		
System version	Version label	Intensity (0.200000)
1	Alpha	-
2	Alpha2	-
3	alpha3	0.027027
4	Beta1	0.003906
5	Release	0.001942

- About the failure intensity :
- This number is the ratio between the number of found defects and the number of test steps executed without defects.
  This number should normally decrease. Examinating the curve could be helpful to predict the time before a release version.

Back to main

Figure 2.36. *Indicator of failure intensity*

2.4. Examples of industrial applications

2.4.1. *AUDI*

AUDI is a premium car manufacturer and a subsidiary of Volkswagen based in Germany.

2.4.1.1. *Modeled systems*

AUDI used the MaTeLo tool to model many functionalities, among which were Start & Stop, Safety computer, additional heating, air-conditioning, integral lighting, or various engine control functions, which are validated in the phases called computer validation or functional integration test.

2.4.1.2. *The equipped test chain*

The test automation environment is based on the free solution EXAM (developed by MicroNova for Volkswagen), which controls HIL dSpace benches, among others.

Managing the requirements is carried out with the Rational DOORS tool. The design of the tests is carried out with the MaTeLo tool.

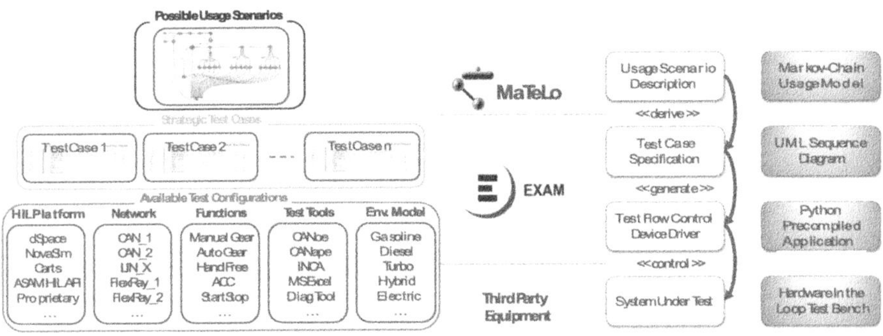

Figure 2.37. *The equipped test chain at AUDI*

2.4.1.3. *Return on investment*

Modeling the tests makes it possible for AUDI to formalize its test specifications and to obtain the following:

– a clear separation between modeling and programming;

– results of analysis work of the requirements available via models, which are better structured, more comprehensible, and more easily maintainable;

– systematic improvement of its specification quality thanks to the detection of errors ahead of the phase;

– an optimized output of its test strategy by choosing the most relevant test cases.

In a pragmatic way, MaTeLo now allows the tests' originators to formalize in the test model what they had only mentally formalized before with a traditional approach. Moreover, formalizing all the scenarios enables them to identify what is tested, but also what is not. The project leader is thus better able to ensure a good test strategy.

2.4.2. *Magneti marelli*

Magneti Marelli is a high-end automobile equipment supplier, a subsidiary of FIAT and the leader in dashboard design and manufacture based in France. Magneti Marelli was one of the leading partners of the European MaTeLo project directed by ALL4TEC. It was the first to use MaTeLo in a productive and industrial way.

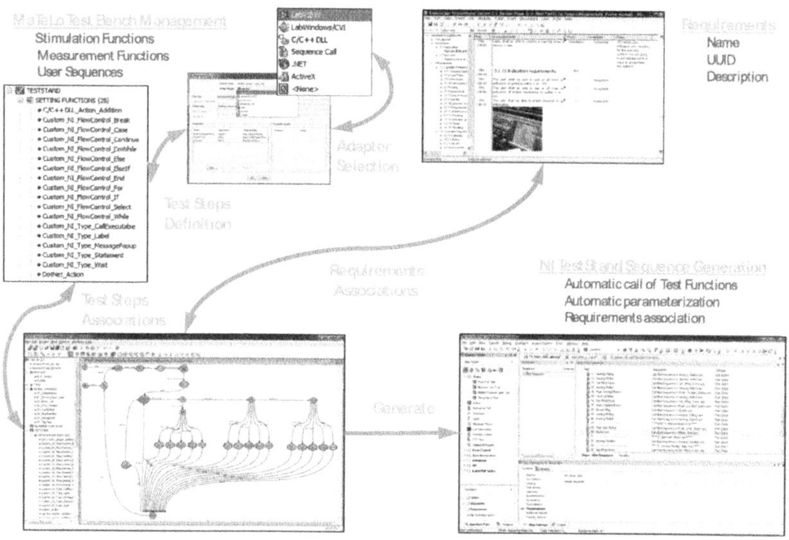

Figure 2.38. *The equipped test chain at magneti marelli*

2.4.2.1. *Modeled systems*

Magneti Marelli Châtellerault systematically uses MaTeLo to model dashboard testing designed for various European brands.

2.4.2.2. *The equipped test chain*

The test automation environment is based on National Instruments' solutions: in particular TestStand, LabWindows/CVI, LabView and LabView Vision.

Managing the requirements is carried out with the Rational DOORS tool. The design of the tests is executed systematically with the MaTeLo tool.

2.4.2.3. *Return on investment*

The use of the MaTeLo tool and its algorithm "Arc Coverage" makes it possible for Magneti Marelli to systematically ensure a 100% functional coverage (100% of the transitions tested), even in the event of specification modification, while obtaining reduced times. The *User-Oriented* algorithms make it possible to ensure the reliability of the dashboards for their various future uses.

In a pragmatic way, the test design is done simultaneously with that of the solution in an iterative fashion. It is thus easy to validate each function from its delivery, but also to check, optimize, and gauge (OCR training) the bench tests in a progressive manner. With every iteration, the function is individually validated, and then tested in interaction with the functions previously validated until the dashboard is completely validated.

2.4.3. *Other industrial applications*

Without going into the details such as with AUDI and Magneti Marelli, we can say that the principles of test modeling supported by the MaTeLo tool are used in many fields and for programmed systems of all types.

For example, in the following fields:

– automotive: validation of body computers, sliding side doors, air-conditioning, adaptive functions for regulating speed, hands-free access, managing memory drivers, etc.;

– railway: validation of management for the corridor doors, or TCMS (*Train Control Management System*), etc.;

– industry: testing autonomous systems of electric transformers and supply, military equipment, or house automation, etc.;

– information systems: MMI testing of communication systems, telephony management systems in business, or supervision and issuing of transport tickets, etc.

2.4.4. *Industrialization of the tests*

In a general way, all the projects of evaluation, leaders, or industrialists showed interest in the MaTeLo solution from the modeling phases of the usage scenarios. The discovery of errors or inconsistencies in the specifications is systematic like the discovery of defects during the execution of tests.

Modeling allows the originators of the tests to raise relevant questions to understand the expected behavior at the system level, but also to understand how to test it.

The models formalized under MaTeLo become supports of exchange and dialogue between the persons in charge of the need and solution of tests for a more formal and coherent collective comprehension.

The various possible strategies allow them to effectively direct the tests while profiting from more diversified and encompassing test cases and providing enough tests to maximally load the automatic test environments.

It is now possible to industrialize the test process from the requirements capture, through the usage model definition and test cases generation, until the test results' exploitation in a full integrated and automated way. The methods described previously make it possible to improve testing throughout the project and thus to limit the individual errors and support group decisions.

2.5. Conclusion

Using models to carry out the phases of a functional test in complex systems is in full expansion.

Solid theoretical bases exist and tools like MaTeLo make it possible to implement them in the industrial world.

MaTeLo made it possible for its various industrial users to make considerable profits despite the test costs and for the testers to devote themselves further to test strategy and its effectiveness so that implementation becomes much less laborious.

MaTelo continues to evolve since marketing its first version in 2004. It has now reached a completely satisfactory level of maturity, but the company ALL4TEC always devotes a supported investment in related R&D.

The prospects of the tool's planned evolution are:

– *in the short term*:

- improving collaborative modeling,
- a new open structure based on Eclipse,
- new-generation algorithms;

– *in the mid-term*:

- modeling and validation of projects managed in product lines,

- generating test campaigns based on risk,

- integrating concepts of estimated reliability allowing the remaining test load to be calculated according to quality criteria;

– *in the long term*:

- formal verification of usage models thanks to the creation of a language user,

- automatically accounting for specification modifications in test strategies.

2.6. Bibliography

[APP 95] APPLEGATE D., BIXBY R., CHVATAL V., COOK W., Finding cuts in the TSP: a preliminary report, Technical Report, 1995.

[DAN 54] DANTZIG G.B., FULKERSON R., JOHNSON S.M., *Solution of a Large-Scale Traveling Salesman Problem*, Operations Research, 1954.

[DYE 92] DYER M., *The Cleanroom Approach to Quality Software Development*, John Wiley and Sons, London, 1992.

[GOS 04] GOSEVA-POPSTOJANOVA K., KAMAVARAM S., "Software reliability estimation under uncertainty: generalization of the method of moments", *Proceedings of the 8th IEEE International Symposium on High Assurance Systems Engineering*, Florida, March 2004.

[JON 08] JONES C., *Estimating Software Costs*, McGraw Hill, New York, 2008.

[LEG 05] LE GUEN H., Validation d'un logiciel par le test statistique d'usage: de la modélisation à la décision de livraison, PhD thesis, Rennes 1, 2005.

[MUS 99] MUSA J.D., *Software Reliability Engineering*, McGraw-Hill, London, 1999.

[NTA 98] NTAFOS S., "On random and partition testing", *International Symposium on Software Testing and Analysis*, Florida, March 1998.

[OST 88] OSTRAND T.J., BALCER M.J., "The category-partition method for specifying and generating functional tests", *Communications of ACM*, June 1988.

[PET 00] PETRENKO A., "Fault model-driven test derivation from finite state models", *Proceedings of Modeling and Verification of Parallel Processes (MOVEP 2000)*, Nantes, 2000.

[SAY 99] SAYRE K., Improved Techniques for Software Testing Based on Markov Chain Usage Models, PhD thesis, University of Tennessee, 1999.

[THÉ 93] THÉVENOD-FOSSE P., WAESELYNCK H., "Statemate applied to statistical software testing", *International Symposium on Software Testing and Analysis*, 1993.

[TRI 82] TRIVEDI K.S., *Probability and Statistics with Reliability, Queuing and Computer Science Applications*, Prentice Hall, Upper Saddle River, 1982.

[VAL 00] VALLÉE F., VERNOS D., "Le test et la fiabilité du logiciel sont-ils antinomiques ?", *Congrès Lambda Mu 12*, 2000.

[VAL 04] VALLÉE F., "La fiabilité des logiciels", *Sciences et Techniques de l'ingénieur, SE 2520*, 2004.

[WHI 92] WHITTAKER J.A., Software testing and reliability analysis, PhD thesis, University of Tennessee, May 1992.

Chapter 3

Safety Analysis of the Embedded Systems with the AltaRica Approach

3.1. Introduction

In this chapter, we describe the use of formal methods to assess the safety of embedded aeronautical systems. We are particularly interested in the AltaRica language and the tools associated with it.

We begin with the presentation of the context of the use of AltaRica: the development of aeronautical systems and the safety assessment activities, which are conducted for certifying aircraft airworthiness.

Then, we introduce the AltaRica language and the associated assessment tools. We show how they were used to assess the safety of two types of aircraft systems. Lastly, we compare our work with other similar approaches and conclude with an assessment of some work and prospects.

3.2. Safety analysis of embedded systems

The certification of the aircraft requires following a process of strict development including safety analyses applied at various stages of an aircraft's development. The three stages usually considered include:

Chapter written by Pierre BIEBER and Christel SEGUIN.

– the development of the functions that the aircraft must offer to its customers (for example, controlling the speed of the aircraft in flight and on the ground);

– the development of the embedded systems that fulfill these functions (such as the thrust reverser deployment system that contributes to slowing down the aircraft on the ground);

– the development of the equipment (such as a computer hosting the software that generates the deployment command of the thrust reversers).

The applicable analyses of safety are not the same ones for the development of systems and data-processing equipment. Different documents describe these analyses. Refer to ARP 4754 [SAE 10] for the analyses applicable to aircraft development and its systems whereas the documents DO 178 version B [RTC 92] and DO 254[RTC 00] relate to the development of data-processing equipment. In the aeronautical field, system development is traditionally carried out by the aircraft manufacturer whereas equipment development is generally carried out by subcontractors. This justifies that document ARP 4754 [SAE 10] is focused on the preliminary aspects of the implementation of the equipment, such as risk analysis, the development of the structure of the system, the establishment of requirements, or on later aspects of implementation, such as checking the integration of the equipment.

In this chapter, we discuss how to use models formally described with the AltaRica language to support safety analyses, which must be carried out when designing the structure of the embedded system. These analyses aim at establishing that the system structure meets the safety requirements allocated to it.

The safety requirements are in the following form:

> "The failure condition FC belongs to a class of severity. The rate of the occurrence of FC must be lower than a quantitative objective. No combination meeting a qualitative objective must lead to FC".

A failure condition (noted FC) is a combination of failure modes applied to the functions allocated to the studied system. The failure modes usually taken into account are the total or partial loss of a function, the detected or undetected erroneous behavior of a function, and the inadvertent activation of a function. The failure condition can also refer to conditions, which describe the current phase of flight (taxiing, takeoff, cruising, etc.), a system operating mode (engine 1 off), or an external condition (crosswind, contaminated runway, etc.).

The classes of severity for safety are *Catastrophic, Hazardous, Major, Minor,* and *NSE* (*No Safety Effect*). During risk analysis, the failure conditions are classified according to their effect on the aircraft and its occupants. The quantitative objective associated with a failure condition is a limit that the rate of occurrence (measured per hour of flight) should not be exceeded. This objective depends on the class of severity of the failure condition.

The usual values are 10^{-9} (10^{-7} for hazardous and 10^{-5} for major) for a condition classified *Catastrophic*. The qualitative objective describes the type of combination of failures, which should not lead to the failure condition. They are most usually simple, double, or triple combinations.

The assessment of the capacity of an aircraft system to maintain the qualitative or quantitative safety objectives is done today mainly using fault-trees or reliability block diagrams. The analysis is done while considering the failure conditions one by one. For each one, the reliability expert builds a tree or a diagram, which gradually breaks down the failure condition into more elementary states until basic failures are obtained (leaves of the tree).

This type of model is, in fact, a Boolean formula whose literals (leaves of the tree) are labeled with probabilities. The associated tools make it possible to calculate the minimal combinations of the literals (or minimal cut sets) which cause the initial failure condition. The analysis of the order of these minimal cut sets (the number of literals intervening in a combination) makes it possible to easily determine if the qualitative objectives are met. The probability of the occurrence of the initial failure condition based on the literal probabilities can also be calculated effectively, starting from the minimal cut sets or directly on the tree.

However, as the systems become more complex, the approach becomes increasingly difficult to control. The number of trees grows with the number of failure conditions to be analyzed. The size of each tree grows with the size of the systems. Furthermore, new systems integrate more and more functions within basic physical components and can be dynamically reconfigurable. Reliability experts need models closer to the system architecture to help share knowledge and monitor the spatial and temporal propagation of failures.

Some languages such as AltaRica aim to overcome these limits while making it possible to build failure propagation models that are dynamic, compositional, and close to architectures and that support the assessments of qualitative and quantitative safety requirements.

3.3. AltaRica language and tools

3.3.1. *The AltaRica language*

The AltaRica language [GRI 98, ARN 00, POI 00] was created by the *Laboratoire Bordelais de recherche en informatique* (LaBRI) (Bordeaux Laboratory for Computer Science Research) at the University of Bordeaux. It makes it possible to describe the behavior of a system in a nominal case and in the presence of failures at the same time. We retained it because it is a simple formal language that is also hierarchical and compositional. Its clearly defined semantics and syntax enable it to be coupled with various reliability and validation tools, such as Aralia ([RAU 93, DUT 97]), Moca-RP, or model-checkers like Mec V ([VIN 03, VIN 04]) or SMV [McM 94]. Continuing in this section, we present the principal components of the fragments of the AltaRica language that we used (AltaRica Dataflow).

3.3.1.1. *AltaRica node*

Each component is described by a `node` that comprises three parts: the declaration of variables and events; the definition of transitions; and the definition of assertions.

Figure 3.1. *Graphic representation of the node Function2*

We illustrate these concepts with the example of a computation function with two input flows. The component *Function2* has two inputs I_1 and I_2 and an output O. When it does not undergo any failure, this function produces a correct value provided that each one of its inputs is correct; it does not produce any result when at least one of these two inputs is absent and it produces an erroneous value in the other cases.

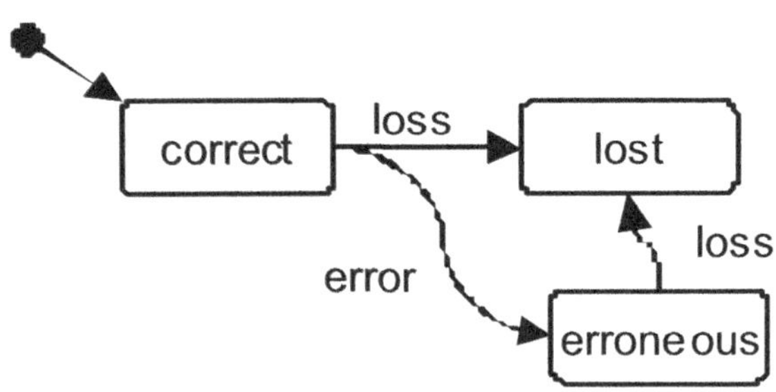

Figure 3.2. *Automaton of the failure modes of the node* Function2

If the node *Function2* is not lost, a failure categorized as an error can occur, which causes the erroneous state. In this state, the node *Function2* produces an `erroneous` value. In the same way, a failure categorized as lost can occur, which causes the state `lost`. In this state, the node *Function2* does not produce a value.

The AltaRica code of the node *Function2* is presented in Figure 3.3.

```
node Function2
      flow
            O: FailureType:out;
            I1, I2: FailureType:in;
      state
            Status: FailureType;
      event
            loss,error
      trans
      (Status != lost) |- loss -> Status:= lost;
      (Status = ok) |- error -> Status:= erroneous;
      assert
      O = case { Status=ok and I1=ok and I2=ok : ok,
            Status=ok and (I1=lost or I2=lost) : lost,
            Status=lost: lost,
            else erroneous};
      init
            Status:= ok;
      extern
            law (<event loss>) = "exp 1e-4";
            law (<event error>) = "exp 1e-5";
edon
```

Figure 3.3. *AltaRica code of the node Function2*

An AltaRica node can have state variables that make it possible to store values; they are declared under the heading `state`. The node *Function2* has one state variable called `Status` whose value indicates the component's functional or dysfunctional mode.

An AltaRica node can also have flow variables which establish the link between the node and its environment; they are declared under the heading `flow`. It is also necessary to indicate the direction of the flow variables: `in` for the input, `out` for the output, `local` for the variables, which are used as abbreviations.

The node *Function2* has two input flows (`I1` and `I2`) and one output flow (`O`). In the models we carried out, these variables are either Boolean or they take their values in a small enumerated set of constants. The four variables of the node *Function2* take their values in a predefined domain called `FailureType`, which is equal to the enumeration `{ok, lost, erroneous}`.

A node can also have `events` that can change the value of the state variables The events are useful to model the occurrence of failures or the reaction to conditions on the flow variables. We indicate if the event is started internally by the component as in the case of a reconfiguration (in this case, the `Dirac` law of probability is associated with the event under the heading `extern`), or externally by its environment as in the case of the occurrence of a failure (in this case, a law of probability of occurrence of the failure as `exponential` is associated with the event under the heading `extern`).

The node *Function2* has two events, which represent failures: `loss`, which corresponds to the loss of this component and `error`, which corresponds to the erroneous behavior.

The `trans` section contains the definitions of transitions being formed:

g |- evt -> e

The guard `g` is a Boolean formula that can refer to the state and flow variables. The guard defines the configuration in which the transition can be carried out if the event `evt` occurs. The effect `e` modifies the value of the state variables if the guard is satisfied and if the event occurs.

The transition associated with the event `loss` can be triggered only if the node *Function2* is not already in the state `lost`. The new value of `state` becomes `lost`. The value of a state variable can be modified only by one transition. The transition associated with the event `error` can be triggered if the node *Function2* is in the state `ok`. The new value of `state` becomes `error`.

The dependency relation between the state of the component and the value of its flows are defined by the assertions (`assert`).

In the models that we made, the assertions describe the calculation rules of the value of output flows according to the state of the component and the value of input flows:

– the assertion associated with the output variable of *Function2* means that a correct value is produced (`O = ok`) when `Status` and the two input flows are equal to `ok`, no value is produced (`O=lost`) when `Status` is equal to `ok` and one of the two input flows is equal to `lost` or;

– no value is produced (`O=lost`) when Status is equal to `lost` (otherwise an erroneous value is produced (`O=erroneous`) (`Status = erroneous`) or `Status=ok` and one of the input flows is equal to `erroneous` and the other is different from `lost`).

The initial value of the state variables is defined under the heading `init`. Initially, `Status` is equal to `ok`. This means that at system startup, the component *Function2* is supposed to function correctly.

3.3.1.2. *AltaRica model*

In a system model, the instances of AltaRica nodes are interconnected by assertions that connect input flows of a component to output flows of other components. Figure 3.4 shows a model made up of an instance of the node *Function2* named `fc` and of two instances of the node *Function0* named `s1` and `s2`. The output of `s1` and `s2`, is connected to the input of `fc`.

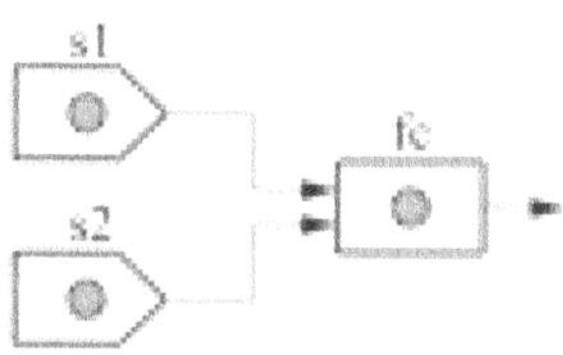

Figure 3.4. *Graphic view of an AltaRica model using Function2*

The AltaRica code presented in Figure 3.5 contains the AltaRica code of the two nodes used (*Function0* and *Function2*) as well as a principal node (`main`).

The heading `sub` of the principal node makes it possible to declare the two instances of *Function0* (`s1` and `s2`) and the instance of *Function2* (`FC`).

The assertions express the fact that at any time the value of the output `O` of `s1` (respectively of `s2`) should be equal to the input value `I1` (respectively `I2`) of `fc`.

Although directly linking flow variables is the most used method of describing the interactions of AltaRica nodes, it is also possible to model the interactions by combining the events of different nodes.

For each group of events, a new event representing this group is added to the AltaRica principal node. There are two types of combining events:

– *synchronization event*: the synchronization event can be triggered if the combined events can be triggered individually (the transition guards of these events are true). When a synchronization event is triggered, all the combined events are triggered and all the variables are modified in accordance with the defined transitions;

– *broadcast event*: the broadcast event can be triggered if at least one of the combined events can be triggered on its own. When a broadcast event is triggered, all the combined events whose guard is true are triggered and only the variables corresponding to these transitions are modified.

```
node Function0
      flow
            O: FailureType:out;
      state
            Status: FailureType;
      event
            loss,error
      trans
      (Status != lost) |- loss -> Status:= lost;
      (Status = ok) |- error -> Status:= erroneous;
      assert
      O = Status ;
      init
            Status:= ok;
      extern
            law (<event loss>) = "exp 1e-4";
            law (<event error>) = "exp 1e-5";
edon

node Function2
      // code for Function2 deleted
edon

node main
      sub
            s1,s2: Function0,
            fc: Function2;
      assert
            fc.In1 = s1.O,
            fc.In2 = s2.O;
edon
```

Figure 3.5. *AltaRica code of a model using Function2*

In both cases, it is possible to decide if the events belonging to the groups must be hidden. A hidden event cannot be triggered individually.

3.3.1.3. *Semantics of AltaRica*

The article [ARN 00] defines the semantics of AltaRica mathematically. In this section, we informally explain the semantics of the Dataflow fragment used.

An AltaRica node describes a mode automaton in which each state is a configuration, i.e. a function that associates a value to each flow and state variables.

The initial configurations are defined by the values of the state variables which appear in the clause `init`. There are as many initial configurations as acceptable combinations of the values of the input variables.

Once the values of the input and output variables are fixed, the values are associated with the outputs after assessment assertions.

In the automaton associated with the AltaRica node, two configurations are connected by an arrow labeled with an event name if, in the starting configuration, the guard of the transition associated with this event is true and the state variables of the arrival configuration are updated in accordance with the assignment of the transition. If the variables do not appear in the assignment, they keep the value assigned to them in the starting configuration.

A scenario of model events is a path in the automaton that starts in the initial configuration and goes from configuration to configuration by selecting an arrow labeled by an event. If, in the current configuration, there is an arrow labeled by an event whose law of probability is `Dirac(0)`, then all the events whose law of probability is not `Dirac(0)` are not selectable in this configuration. This means that the `Dirac(0)` events have priority compared to the other events.

In the models that we built, we use `Dirac(0)` events to represent the actions of automatic reconfiguration whereas we use the other events to represent the equipment failures. Thus, in these models, the reconfiguration events are always triggered before the next possible failure.

An AltaRica model made up of several interconnected nodes can be represented by a single node containing the set of variables, events, assertions, and transitions from the contained nodes. The rules of semantics described previously apply to the whole node, which makes it possible to describe the behavior of the complete AltaRica model. The AltaRica models, which we designed do not contain free input variables (all the input variables are connected to an output variable). Consequently, there is a single initial configuration for this type of model.

3.3.2. *Modeling the propagation of failures with AltaRica*

We use the AltaRica language to model the propagation of failures. These models [McD 94, HEI 01], describe the failure modes that can occur for each component of a system. They also describe the way in which these failure modes are propagated within the architecture of the system.

Based on the description of a system, the analyst initially defines a library of useful AltaRica nodes to model the system. For example, in the case of an electrical power system, we created a library containing nodes describing the usual equipment, such as a generator, an electric bar, a switch, a circuit breaker, etc.

We also built a useful library of nodes in order to model systems in a more abstract way, for example, in the form of a functional architecture. In this case, the library contains nodes such as *Function0* and *Function2* presented in the preceding section. Each node represents the anticipated failure modes of the modeled component. In the case of an electrical component, it can relate to the loss of output voltage or a short-circuit. In the case of a functional component, the failure modes considered could be the loss of the function, providing an erroneous result, unnecessary operation, etc.

Each failure mode is represented by an event node. Moreover, by defining the possible values of the output according to the input values and the status of a component, the assertions describe the propagation of failures within the component. To facilitate modeling the failure propagation, the values of a variable indicate its status from the safety stand point, like an incorrect command or a loss of hydraulic power, rather than the real value that this variable can take.

3.3.2.1. *Description of the "Terrain Following/Terrain Avoidance" system in AltaRica*

The *Terrain Following/Terrain Avoidance* system (TFTA) provides either the flight control system or the pilot of an aircraft with orders to climb or descend so that the aircraft maintains a proper altitude compared to the ground. This system makes it possible for a military aircraft to fly very close to the ground.

Figure 3.6 shows the principal functions and the data flows of the terrain following system. The rectangles represent the functions and the hexagons represent the data flows.

Before using TF/TA, the pilot selects the height *SelHeight* of the aircraft by using a dedicated interface called the *TF/TA panel*. Then, a vertical acceleration *VertAccel* is computed with the following input flows: *SelHeight* (previously defined), *TerrainInfo* (information on the relief provided by the radar), and the speed of the aircraft (computed by the navigation system). Then *VertAccel* is sent to the flight control system.

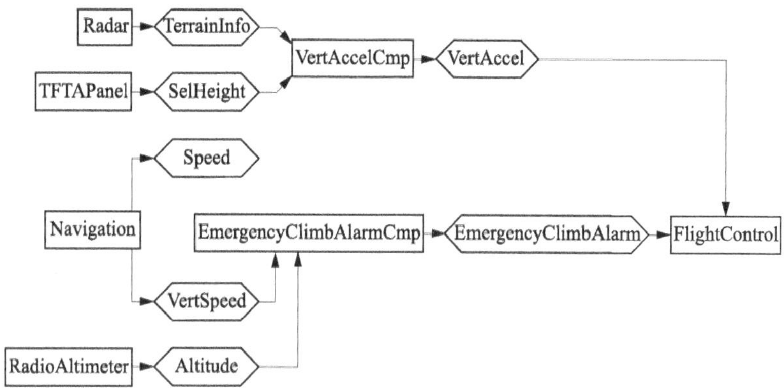

Figure 3.6. *The terrain-following system*

Simultaneously, a release alarm *EmergencyClimbAlarm* is calculated on the basis of the vertical speed provided by the navigation system and the altitude provided by the radio altimeter. This alarm is sent to the flight control system so that the aircraft can go up quickly and reach a safe altitude.

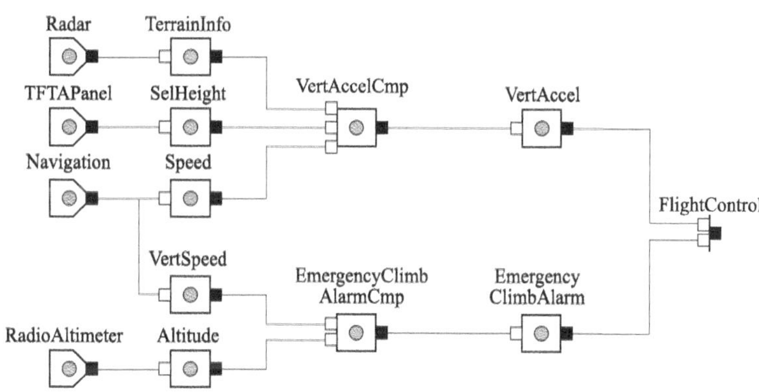

Figure 3.7. *AltaRica model of the terrain-following system*

The AltaRica model presented in Figure 3.7 was derived from the description of the terrain-following system. The model is built using instances of AltaRica nodes, which belong to a library including the *Function0* and *Function2* nodes presented earlier. This library contains also the node *Function3* (respectively *Function1*) which is similar to *Function2* with one more (respectively less) input flow.

The names of the instances and their interconnections are directly extracted from the functional description of the terrain-following system. An instance of the node *Function2* is associated with the function *EmerClimb AlarmCmp*. An instance of the node *Function3* is associated with the function *verAccelCmp*. An instance of the node *Function1* is associated with each data flow. Finally, an instance of the node *Function0* is associated with all the other functions except *FlightControl*.

3.3.2.2. *Modeling the requirements*

As previously described, a safety requirement is related to a failure condition FC, which is a combination of couples <functions, failure modes>. To represent the safety requirements, nodes named "observers" are added to the model. These nodes have input variables that make it possible to observe the status of certain components. An observer calculates the value of output variables that indicate the status in which a failure condition occurs.

The safety requirements of the terrain-following system are related to three failure conditions:

– *terrain-following produces an undetected erroneous vertical acceleration* (TFTA_*erroneous*), this situation is extremely critical because it could lead to the aircraft crashing. In the AltaRica model, this situation occurs when the output variable `VertAccel. O` is equal to `erroneous` and that `EmergencyClimbAlarm. O` is not equal to `ok`;

– *terrain-following is lost* (TFTA_*loss*). This situation is supposed to be less dangerous than the preceding because the pilot can stop terrain following and fly manually. Nevertheless, this situation can involve canceling a mission, which would require using terrain following. In the AltaRica model, this situation occurs when the output variables `VertAccel.O` and `EmergencyClimbAlarm.O` are both equal to `lost`;

– *the terrain-following system produces a spurious release alarm* (TFTA_*spurious*). This situation is not dangerous because the aircraft will carry out a release operation that will make it reach a safe altitude. Again, this situation can result in canceling a mission that necessitates terrain following. In the AltaRica model, this situation occurs when the output variable `VertAccel.O` is equal to `ok` and `EmergencyClimbAlarm.O` is not equal to `ok`.

The node called *FlightControl* is an observer that makes it possible to calculate if one of the three failure conditions just presented has occurred. To carry out this calculation, *FlightControl* takes the input and output of the data flows *VertAccel* and *EmergencyClimAlarm*. The safety requirements associated with the terrain following system are summarized in Table 3.1.

FC	Safety	Qual.	Quant.
erroneous	Haz	2	1e-7
loss	Maj	2	1e-5
spurious	NSE	1	1e-3

Table 3.1. *Safety requirements of terrain following*

The first line of the table indicates that the production of an undetected erroneous vertical acceleration should be caused only by combinations of at least two failures and that the rate of occurrence of this event must be lower than 10^{-7} per hour of flight.

The second line indicates that the loss of terrain following should be caused only by combinations of at least two failures and that the rate of occurrence of this event must be lower than 10^{-5} per hour of flight.

The last line indicates that the production of a spurious alarm does not have an associated qualitative requirement but that its rate of occurrence must be lower than 10^{-3} per hour of flight.

3.3.3. *Tools associated with AltaRica*

The development of AltaRica models is supported by industrial and academic tools, which provide functionalities such as a graphic editor for models and a components manager. Moreover, these tools also offer functions facilitating safety analysis. In this section, we present in more detail the various tools allowing AltaRica model analysis.

3.3.3.1. *Interactive graphic simulator*

The interactive graphic simulator is an extremely useful tool for understanding the behavior of a model. The tool graphically presents the AltaRica model as well as its current configuration. The color of the lines between the nodes indicates the current value associated with the input and output variables connected by these lines.

Icons are associated with each state of a node. For example, a green box is posted when the state variable of the component is equal to `ok`, a red box is posted when this variable is equal to `lost`, and a blue box is posted in other cases. The icons and the colors of the links make it possible to quickly assess the current state of a set of components. It is also possible to inspect in detail the values of the model variables by using a textual interface, which lists the set of the variables and their current value.

First, the simulator calculates the initial configuration of the model. Then, when the analyst selects a node, the simulator proposes the set of events that are able to be triggered in the current state. This relates to the set of events whose guard is true.

The analyst chooses an event and the resulting state is calculated by the simulator. As the majority of the events of an AltaRica model represent failures, this allows the analyst to insert several failures into the model in order to observe whether a failure condition has occurred. By using simulation, it is possible to validate a model with a system design team.

It is wise to first examine the behavior of each AltaRica node to make sure that the failure modes of the component were correctly taken into account in the modeling.

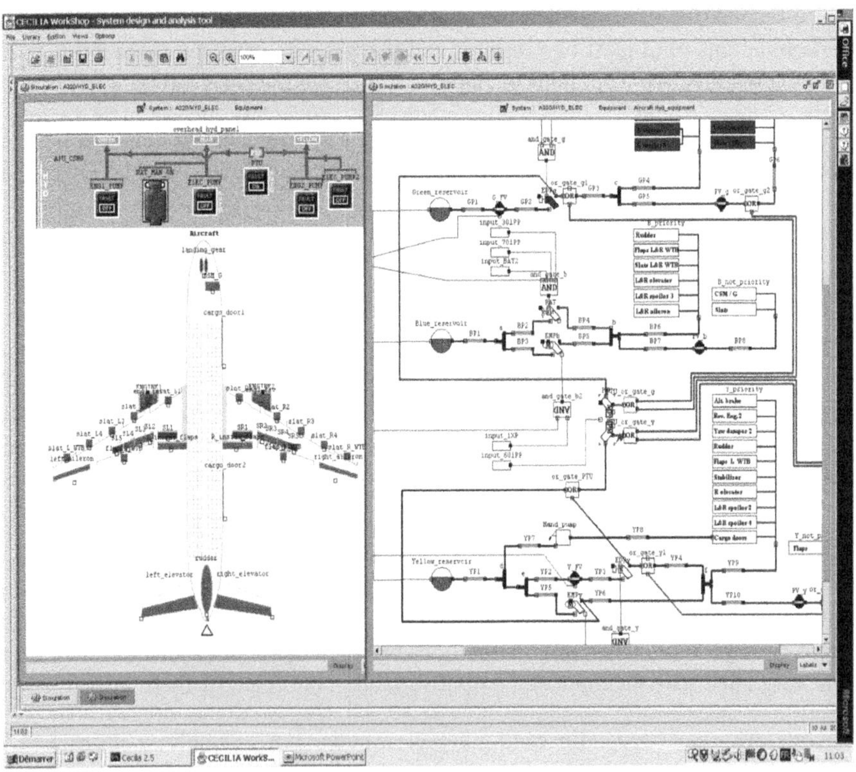

Figure 3.8. *Graphic simulator of Cecilia OCAS*

Then, it is necessary to observe the model to determine whether the composition of the instances of nodes represents the propagation of the failures correctly. It is a question of making sure that the propagation modeling of the failures is consistent with the behavior of the real system. For that, the system designers can define standard scenarios of behavior, which will be played-out with the interactive simulator to check whether the states reached by the simulator are identical to those envisaged by the scenarios.

Figure 3.8 presents the interface of Dassault Aviation's Cecilia OCAS workshop. The system model of hydraulic power generation and distribution is located in the right-hand side window. Each icon represents a component (tank, pump, valve, gate, etc.) of this system.

The window on the left shows a set of observant nodes that indicates if the aircraft equipment, such as the servo valves on mobile surfaces (slat, flaps, ailerons, etc.), the brakes, or the thrust reversers receive hydraulic power. At the top of this window, there is a set of components representing the devices, which the pilot uses to activate and deactivate the hydraulic system equipment.

3.3.3.2. *Tools of analysis*

Several tools can be used to check whether the AltaRica model of a system meets the expected safety requirements.

The first group of tools generates the minimal combinations of failures that lead to a failure condition. The size of these combinations is analyzed to determine if the qualitative requirements are satisfied. The combinations can also be provided to tools carrying out probabilistic calculations to determine if the quantitative requirements are satisfied.

There are several means of generating the minimal failure combinations. First, fault tree generation [RAU 02] allows, for certain AltaRica models, automatic production of a fault-tree. This tree represents a Boolean formula whose variables are the events of the AltaRica model. The formula describes the paths that lead the system from the initial state to a state that satisfies the failure condition. The minimal cuts are calculated starting from the fault-trees using a tool such as Aralia. Another approach consists of using the AltaRica model simulator to directly generate a set of minimal sequences that are smaller than a fixed limit and that lead to the failure condition.

The principal difference between the two approaches is that a sequence represents the exact chronology of the events whereas a minimal cut is a set of events whose order of appearance is not important.

The second difference between these approaches relates to their limitations. Generating a fault-tree is usable only for static models. In these models all the permutations of a set of events lead to the same state. When this property is not checked by the model, we talk about a dynamic model; it is necessary to use sequence generation. The generation of sequences truncates the set of provided results because the search for sequences is limited by a maximum sequence size.

It is thus possible that sequences larger than the limit exist. This does not impact the behavior analysis of the qualitative requirements since we are interested in sequences smaller than the limit. On the other hand, this can impact the calculated probability because these truncated sequences could increase the probability of occurrence of the failure condition. Because of computing times, it is not always possible to increase the limit freely because the number of sequences to be examined grows quickly according to the number of events in the model and the limit. When a fault-tree can be generated, the available tools can extract all the minimal cut sets without limiting their size. This approach is consequently likely to provide more precise quantitative results than the sequence generation.

The terrain following system model presented previously is a static model. Consequently, it is possible to generate a fault-tree and to calculate the minimal cuts for the three failure conditions (*TFTA_loss*, *TFTA_erroneous*, and *TFTA_spurious*). The minimal cuts are presented in textual form, each line giving a combination of failures leading to the failure condition.

Figure 3.9 shows some minimal cuts generated for *TFTA_loss*. There is one combination including one failure (the loss of the *navigation* function) and 73 including two failures (for example, the error of the navigation function combined with the loss of data flow conveying vertical acceleration) that lead to the loss of the terrain-following system.

```
{Navigation.fail_loss}
{Navigation.fail_erroneous, VertAccel.fail_loss}
{Navigation.fail_erroneous, VertAccelCmp.fail_loss}
{Navigation.fail_erroneous, TerrainInfo.fail_loss}
```

Figure 3.9. *Minimal cuts leading to the loss of terrain-following (extract)*

These first results obtained with the functional terrain-following model show that the qualitative requirements do not seem to be satisfied because the loss of the navigation function leads to the loss of terrain following.

In practice, the qualitative requirement is checked because the navigation function consists of two independent sub-functions, one calculating speed and the other vertical speed. The failure of only one of these two sub-functions does not lead the loss of terrain following.

The set of all the combinations of failure that lead to a failure condition explains why the safety requirements associated with the failure condition are or are not met. Nevertheless, a second group of tools can check whether the safety requirements are met without providing the user with this kind of explanation. These tools are based on the *model-checking* technique. They explore symbolically the automaton associated with an AltaRica model. A *model-checker* can test if a qualitative safety requirement described as a formula of temporal logic is satisfied.

When the formula is not satisfied, the *model-checker* produces a counterexample, which gives the sequence of states that leads to the requirement violation. Several *model-checkers* can be used for checking AltaRica models. MecV, developed at LABRI [VIN 03], offers the advantage of directly treating the AltaRica models. Other *model-checkers* (various alternatives of SMV [McM 93, BOZ 07], and SCADE Prover plugin [ABD 04]) are usable after having used a translator from AltaRica for their input languages [GRI 06].

Finally, we must mention tools that allow users to directly assess the behavior of quantitative requirements. It relates to stochastic simulators or to the analysis of Markov chains [BOI 06, BOU 86].

3.4. Examples of modeling and safety analysis

3.4.1. *Integrated modular avionics architecture*

The embedded systems in modern aircraft, whether they are for military or civilian use, are supported more and more by an avionics platform made up of shared resources for computation and communication.

This new architecture based on shared resources is supposed to have many advantages, like simplifying the installation of the electronic components and their wiring aboard the aircraft or reducing the number of necessary spare parts. Nevertheless, this architecture has an important impact on the development process of embedded systems for two reasons that we have studied:

– a new actor intervenes in the development process. It relates to the team in charge of designing the avionics platform. From now on, the selection of the data-processing equipment is not within the competence of the teams in charge of developing the embedded systems consuming shared resources such as flight control, navigation, or terrain following systems;

– shared resources adds new dependencies between the embedded systems, which are difficult to assess. For example, equipment failure of the shared network could involve the loss or the erroneous behavior of several data flows useful for the systems supported by the platform. Thus, a failure of a single shared piece of equipment could lead to multiple failure conditions implicating several embedded systems.

We propose to use AltaRica to help the designers assess the safety of the systems while accounting for the impact of resource sharing.

We also hope to help the designers of the avionic platform propose a correct hardware architecture, i.e. a physical architecture which can safely support the processing of all the software functions and the data flows of the embedded systems.

In this section, we illustrate the proposed approach by taking the example of the terrain-following system. We are particularly interested in defining an avionics platform likely to host this system.

3.4.1.1. *Description of the avionic platform*

Figure 3.10 describes an avionics platform made up of three computers (Cpu1, Cpu2, and Cpu3) and two buses of communication (Bus1 and Bus2).

In this extremely simple platform, the computers are connected to all the buses. This signifies that they all have the capacity to emit and to receive messages on the two buses of communication.

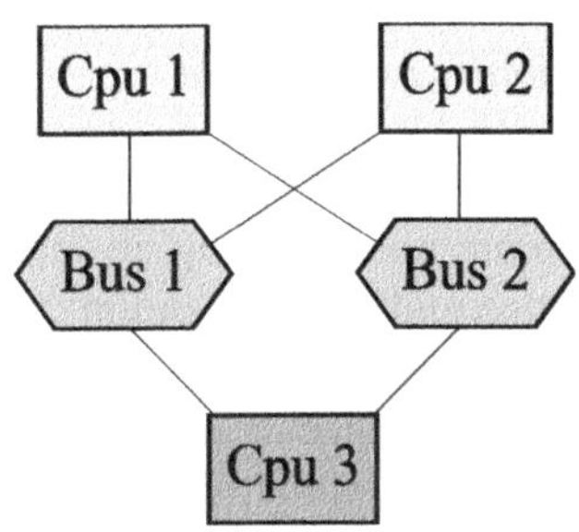

Figure 3.10. *Avionics platform*

Table 3.2 defines two possible allocations of the functions and data flow of the terrain-following system in the avionics platform defined in Figure 3.10.

An allocation is a mapping. It maps functions of the system to computation resources. It also maps data flow to communication resources. It should be noted that a computer can also be regarded as a resource. In effect, when two functions are hosted by the same computer, we assume that the computer allows the functions to communicate directly without using a communication bus.

This explains why the data flow *Altitude* is hosted by computer *Cpu2* because the transmitting and receiving functions of this data flow (*RadioAltimeter* and *EmergencyClimbAlarmCmp*) are hosted by this same computer.

Resource	Allocation 1	Allocation 2
Cpu1	Radar, TFTAPanel, VertAccelCmp, TerrainInfo, SelHeight, Speed, Navigation	Radar, TFTAPanel, VertAccelCmp
Cpu2	RadioAltimeter, Altitude, EmergencyClimbAlarmCmp	RadioAltimeter, Altitude, EmergencyClimbAlarmCmp, VertSpeed, Navigation
Cpu3	FlightControl	FlightControl
Bus1	VertAccel VertSpeed	VertAccel, TerrainInfo, SelHeight, Speed
Bus2	EmergencyClimbAlarm	EmergencyClimbAlarm

Table 3.2. *Table of allocations*

3.4.1.2. *AltaRica model of the avionics platform*

To analyze the impact of resource sharing on the terrain-following system, a new AltaRica model is created. It consists of three parts: the model of the functions and data flow of the terrain following system that was already presented; a model of the avionics platform; and a model of the allocation of resources.

The avionics platform model is carried out by using components in a library containing the nodes representing the CPU, BUS, Gateways equipment that we will not detail. The level of detail for these nodes can be very simple.

In this case, the Function0 behavior presented earlier will be used to model the resources. When the technology used to build the platform is known, it is possible to build more detailed models.

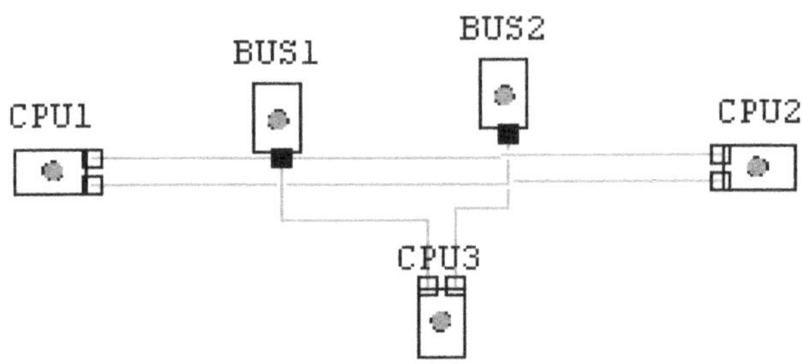

Figure 3.11. *AltaRica model of the avionics platform*

The allocation is modeled by grouping the event that represents the failure of a resource with the events that represent the failures of the functions and data flow hosted by the resource. More precisely, we create for each resource *R* two global events `Rloss` and `Rerroneous`. The event `Rloss` (respectively `Rerroneous`) combines the events `loss` (respectively `erroneous` events) of the functions and data flows hosted by *R* according to the allocations in Table 3.2. For example, the event `CPU1Loss` combines the events `Radar.loss`, `TFTAPanel.loss`, and `VertAccelCmp.loss`.

As we presented in the section devoted to the AltaRica language, a global event can be started when at least one of the combined events (for example, the event representing the failure of the resource) can be triggered. All the combined events (events representing the failure of the functions hosted by the resource) whose guard is true are activated.

Consequently, in the AltaRica model integrating allocation 1, when the global event `Bus1Loss` is triggered, the data flows `VertSpeed` and `VertAccel` as well as the bus `Bus1` are lost.

3.4.1.3. *Safety analysis of the avionics platform*

Table 3.3 summarizes the results of the safety analyses. The first column points out the name of the failure condition considered. The second column gives the type of result reported (*single* for the number of simple failures leading to the failure condition, *double* for the number of double failures, *proba* for the probability of occurrence of the failure condition).

TFTA FC	Result Type	TFTA Function	Global Alloc1	Global Alloc2
Erroneous	*single*	1	2	1
	double	73	0	26
	proba	N/A	2.0 e-5	1.0 e-5
Loss	*single*	1	2	1
	double	77	0	6
	proba	N/A	2.0 e-4	1.0 e-4
Spurious	*single*	10	4	2
	double	0	0	0
	proba	N/A	2.2 e-4	1 e-4

Table 3.3. *Analysis results of the terrain-following system*

The third column gives the results obtained by using only the model of the functions and data flows of terrain following. The two remaining columns provide the results calculated with the global models, which contain the avionics platform as well as the two versions of allocation. The fourth column gives the results obtained with allocation 1 and the fifth column gives the results obtained with allocation 2.

The results obtained with allocation 1 show that this allocation is not acceptable because the simple failure `BUS1Loss` (respectively `BUS1Erroneous`) leads to *TFTA_loss* (respectively *TFTA_erroneous*). Indeed, the loss (respectively the erroneous behavior) of *Bus1* causes the loss (respectively the erroneous behavior) of the data flows *VertAccel* and *VertSpeed.* To make this allocation acceptable, it is necessary to preserve the independence of the data flows *VertAccel* and *VertSpeed.* These two data flows shall not be transmitted by using the same bus of communication.

The second allocation presented in Table 3.2 allocates Bus1 with the data flow *VertSpeed* and computer Cpu2 with the data flow *TerrainInfo*. This contributes to correcting the allocation 1 defect.

The analysis results of the last column in Table 3.3 show that this second allocation is better than the first from the qualitative point of view since it does not add single failure combination compared to the analysis carried out with the model containing only the functional description of terrain following. Allocation 2 is also better than allocation 1 from the quantitative point of view since the calculated probabilities are two times lower with allocation 2.

Nevertheless, it should be noted that these probabilities are not acceptable because they do not respect the objectives laid out in Table 3.2. To correct this problem, it is sufficient to add a new computer Cpu3 and to allocate it exclusively to the function *Navigation*.

3.4.1.4. *Partial conclusion*

The approach of separate modeling of functional and material architectures that have just been presented coheres to industrial practice. Indeed, they are different teams that are in charge of defining the terrain-following system and the avionics platform.

Thus, according to our approach, the model of the functional architecture would be carried out by the designers of the systems whereas the model of material architecture would fall within the competence of the designers of the platform, and the model of allocation could be carried out by the integrator of the systems on the platform.

This type of modeling makes it possible for the designers to easily test several solutions of functional, material, or allocation architectures (see [SAG 05] for the description of an incremental process of system allocation on the platform that rests on AltaRica models).

3.4.2. *System of electric power generation and distribution*

3.4.2.1. *Architecture of the electrical system*

The role of the electrical system is fundamental in an aircraft since it supplies many critical elements (i.e. airfoil servo control, embedded computers, etc.). As the loss of electric power can lead to the loss of aircraft control, this system must respect the following safety requirement: the total loss of electric power is considered catastrophic. The rate of occurrence of this failure condition must be lower than 10^{-9} per hour of flight and single failure should not lead to this failure condition.

To respect this requirement, an electrical system of an aircraft such as the A320[1] is organized in two subsystems: the normal system and the back-up system. Each subsystem includes generators, supply bars, transformers, circuit breakers, converters and junctions.

---

1 To know more about the architecture of the Airbus, see Chapter 6 of [BOU 09].

The normal system is composed of two principal generators GEN1 led by engine 1, GEN2 led by engine 2, and an auxiliary unit of power APU (*Auxiliary Power Unit*). The back-up system is composed of a standby generator CSM_G supplied by the hydraulic system and the RAT (*Ram Air Turbine*) that is automatically deployed if the principal generators are lost.

The electric power is provided to the electric charges by means of four busbars for the nominal system: ACside1, DCside1, ACside2, DCside2, and by means of two essential busbars in the backup system: ACess and DCess. The conversion of the AC current toward the DC current is carried out by transformers TR1, TR2, and TRess. The system also includes circuit breakers to limit the propagation of short-circuits. Contactors are controlled to implement several types of reconfigurations. For example, all the normal system generators (GEN1, GEN2, or APU) can be used to supply electricity to all the busbars of normal or backup systems when one or two generators are not usable.

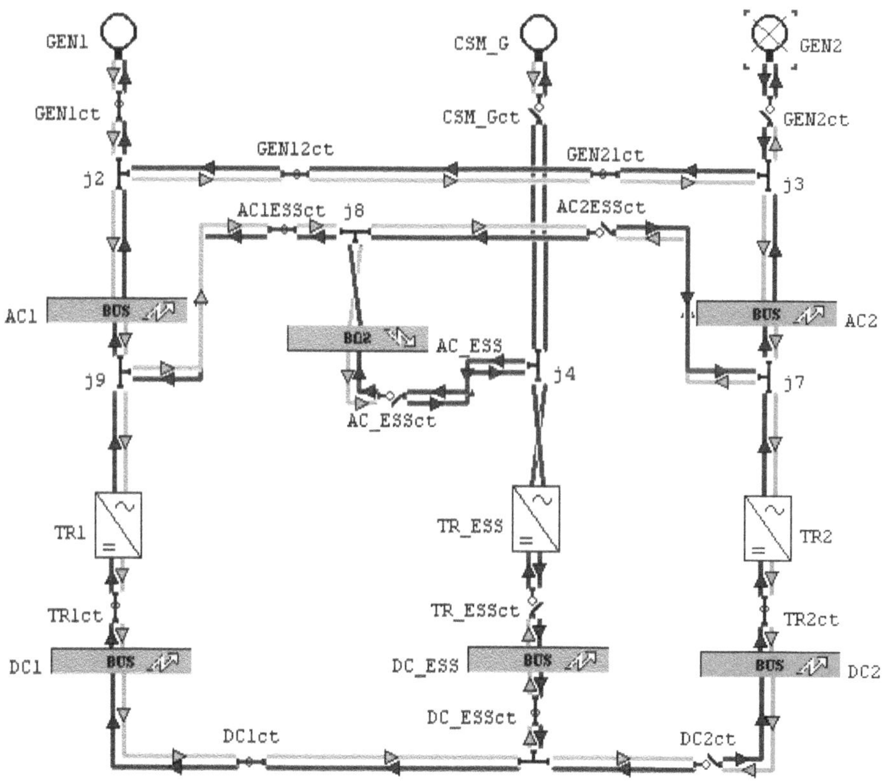

Figure 3.12. *Graphic representation of an electric system model*

Other rules of reconfiguration apply when one of the transformers is lost. Finally, when all the generators of the normal system are lost, the standby generator exclusively provides electricity to the essential busbars (ACess and DCess).

3.4.2.2. *AltaRica model of the electric system*

We developed a library of electric components: generators, circuit breakers, transformers, contactors, busbars, and electrical consumers. For each component, we modeled the effect of the modes of failure on its internal behavior as well as the propagation of failures toward its environment. We were interested in failure modes that could lead to the total or partial loss of electric power.

We assume that all the components could no longer generate, transmit, or provide electricity. In a more detailed version of the model, we had also assumed that short-circuits could occur in the supply bars (see article [KEH 04] for more detail). Finally, the contactors and the circuit breakers can remain blocked in an open or closed position.

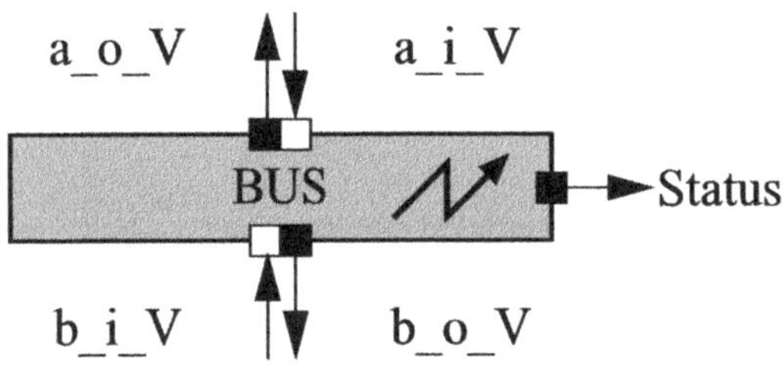

Figure 3.13. *Flows associated with a busbar*

We use the flows of components of the electric system to model the propagation of the failures within the electric system. It is therefore necessary that the flows account for the presence or the absence of voltage. We considered that each port of the component was likely to have an input and output voltage. For example, a busbar with two ports `a` and `b` will be modeled by a node with five flow variables: `a_o_V`, the output voltage in port `a`; `a_i_V`, the input voltage in port `a`; `b_o_V`, the output voltage in port `b`; `b_i_V`, the input voltage in port `b`, and `status` that indicates whether the bar is being supplied.

The system comprises contactors that allow a great number of possible reconfigurations.

We developed controller components that control each of the model's contactors by opening or closing them according to the observation of the current state of the engines, generators, bars, and transformers.

The electric system uses four controllers: GEN for the contactors associated with the generators; AC for the contactors associated with the bars with AC current; TR for the contactors associated with the transformers; and DC for the contactors associated with the bars with DC current (see Figure 3.14).

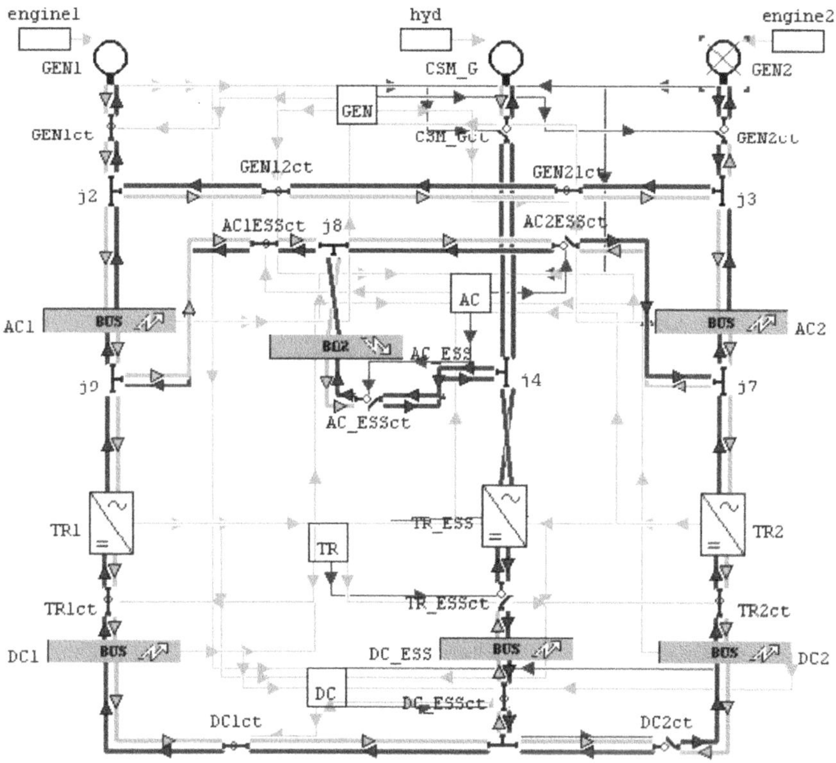

Figure 3.14. *Graphic representation of an electric system (with controllers)*

The AltaRica node in Figure 3.15 describes controller GEN. It has four input flows that describe the status of the engines and bars AC1 and AC2. It has five output flows, one by controlled contactor, which is true when the corresponding contactor must be closed. The rules for reconfiguring the electric system are modeled by logical equations connecting output and input flows.

A first version of the model comprised calculation circularities on the data flows because of physical loops in the architecture of the electric system (i.e. paths, such that a component's input depends on its output). We sought to limit these circularities because the techniques of checking the requirements (model-checker SMV or SCADE, an automatic generator of fault-trees associated with AltaRica) are not capable of treating this type of models.

```
node GEN_Controller
  flow
   ENG1, ENG2 : bool : in ; // status of engine 1, engine 2
   AC1, AC2 : bool : in ; // status of AC1, AC2
   GEN2ct,GEN1ct,GEN12ct,GEN21ct,CSM_Gct: bool : out ; //
   contactor close command
  state
   AC1_fail, AC2_fail: bool ; // stored status of AC1, AC2
  event
   update ;
  init
   AC1_fail, AC2_fail := false, false ;
  trans
   // The update event can occur when AC1 (resp. AC2)
   becomes // not operative, after the update event the
   AC1_fail
   //(resp. AC2_fail) variable reflects that AC1 (resp.
   AC2) is //not operative
   (AC1_fail = false and AC1 = false ) |- update ->
   AC1_fail := true;
   (AC2_fail = false and AC2 = false ) |- update ->
   AC2_fail := true;
  assert
   // GEN12ct : closed when exactly one engine is not
   operative
   GEN12ct = (not ENG1 and ENG2) or (ENG1 and not ENG2);
   //GEN21ct : closed when exactly one engine is not
   operative
   GEN21ct = (not ENG1 and ENG2) or (ENG1 and not ENG2);
   // GEN1ct : closed when engine 1 is operative
   GEN1ct = ENG1 ;
   // GEN2ct : closed when engine 2 is operative
   GEN2ct = ENG2;
   // CSM_Cct : closed when both AC1 and AC2 are unable to
   deliver power
   CSM_Gct = (AC1_fail and AC2_fail);
  extern
   law <event update> = Dirac(0);
edon
```

Figure 3.15. *AltaRica code of a controller model*

One of the loops relates to the control of the contactor `CSH_Gct`. The order to close depends on the state of bars AC1 and AC2, but it also influences the state of these two bars.

To break this type of loop, we introduced a delay into the order to close the contactor.

This is carried out in the preceding model by using the state variables `AC1_fail` and `AC2_fail` that store the preceding state of AC1 and AC2 rather than the flow variables `AC1` and `AC2` that indicate the instantaneous value of these bars. We were thus obliged to add two state variables `AC1_fail` and `AC2_fail` as well as the event `update`, which led these state variables to update. We associate the law `Dirac (0)` to this event, which makes it of a higher priority level compared to all the events representing the failures of the components of the electric system. This makes it possible to guarantee that the reconfigurations of the contactors take place before the next component failure.

3.4.2.3. *Safety analysis of the electric system*

The failure conditions of the electric system are the total loss of electric power (i.e. the loss of three busbars AC or DC), which is considered catastrophic, the loss of two busbars, which is considered major, and the loss of one busbar, which is considered minor. All these requirements are modeled by an *observer* component.

FC	Safety	Qual
Total Loss	CAT	3
Double Loss	MAJ	2
Simple Loss	MIN	1

Table 3.4. *Safety requirements of the electric system*

We could not use the fault-trees generation to analyze the model of the electric system because, to use this tool, it is necessary that the order of appearance of events is not important. This is not checked in the electrical model that we created. Let us consider the two following sequences of events: "loss of the two engines then several `update` of controller GEN" and "several `update` of the controller GEN then loss of the two engines". At the end of the first sequence, the AC bars are supplied whereas in the second sequence, this is not the case because the reconfigurations depending on the event `update` have not been carried out.

To check the requirements, we used sequence generation as well as *model-checking*. Initially, checking the requirements failed, and this was related to deficiencies of the contactor controllers.

We analyzed the generated sequences that invalidated the requirements and corrected the controllers.

The interest in sequence generation is that it is very simple to use and provides a list of minimal sequences leading to the failure condition. Considering that these minimal sequences are comparable to minimal cuts, it becomes possible to assess the model quantitatively. One of the disadvantages of sequence generation is the duration of calculations since they become very long for larger models (model containing more than 200 failures on a mono-processor computer).

In the case where knowledge of the set of sequences is not needed, then *model-checking* offers an interesting alternative. It is enough to check a temporal property indicating that there is no sequence smaller than a certain limit leading to the failure condition. When this check fails, the generated counterexample can be interpreted as a sequence of events leading to the failure condition. This sequence can be played with using the interactive simulator to understand the origin of the problem. This facilitates the correction of the model.

3.4.2.4. *Partial conclusion*

Modeling the electrical system proposed in this section illustrates the principles used for other physical systems (a mechanical part of the flight control device [BER 07], hydraulic system [BIE 04], engine [ADE 10], etc.). Modeling the propagation of the failures requires the use of bidirectional flows; this involves the appearance of circular definitions that must be treated. To eliminate these loops, we proposed adding delays. This makes the model incompatible with fault-tree generation tools. It is thus necessary to use verification tools able to treat dynamic systems, such as sequence generation and *model-checking*.

3.5. Comparison with other approaches

As the preceding sections show, the AltaRica language makes it possible on the one hand to model the propagation of the failures of the components of a system using a static or dynamics, model, reflecting the system architecture and, on the other hand, the associated tools make it possible to use the model in order to qualitatively and quantitatively analyze the occurrence of several failure conditions. Other approaches, more or less recent, were proposed to achieve similar goals.

3.5.1. *Some precursors*

The approaches of FPTN [FEN 94] and HipHops [HEI 01] were proposed to carry out compositional models of the propagation of failures and to extract fault-trees from these models. Each component of the model includes inputs and outputs that conveys failure modes according to a Boolean logic rules specific to the component. These approaches are suitable when the study of the static dependencies between components of the system is possible and sufficient for the needs of the safety analysis.

The networks of stochastic activities [DAL 00] are an extension of the stochastic Petri nets that let the user make models of the propagation of compositional and dynamic failures. The associated Moebius tool makes it possible to estimate various measurements of reliability using stochastic simulations. To our knowledge, the tools for scenario generation applicable to this type of model still need to be developed.

Another language, Figaro, makes it possible to describe a hierarchical system starting from knowledge bases [BOU 91]. According to the type of the knowledge base used, it would be possible to make a static model of the analyzed system and to generate fault-trees. More quantitative analyses on the dynamic part of a system could also be carried out. AltaRica is a less expressive language but was formally defined to model static and dynamic architectures at the same time and to interface itself easily with many existing tools.

3.5.2. *Tools for preexisting formal languages*

More recently, tools for generating fault-trees or sequences leading to a failure condition were developed for compositional formalisms initially conceived to model the dynamics of the complex systems such as SCADE, StateMate, Simulink® – State flow or Nu-SMV [AKE 99, AKE 03]. These languages have *a priori* an expressivity higher than the language AltaRica Data Flow used in this chapter. They thus make it possible to more or less easily carry out models equivalent to those previously presented. However, an AltaRica mode automaton structures the definition of the failure propagation in two parts: one dedicated to expressing the logic of changing the operating mode, and the other dedicated to the logic of dysfunction suitable for a given mode. This is *a priori* simpler than a single automaton which mixes the two parts without distinction, as in former formal languages.

On the other hand, the models written in the richest languages can use functions depending on clocks or on nonlinear functions in their definition of the dynamics of the system. These complex primitives reduce the performances of the current tools considerably and some of them cannot be analyzed at all. Experiments were carried on in the MISSA project ([MIS 11, PAP 11]) for pushing the current limits further and for determining the range of developed tools better.

3.5.3. *Languages for physical systems*

Note that the group of approaches previously cited makes it possible to express functional dependencies (between inputs and outputs of a component and between components) or cause-and-effect relations oriented between the system states. However, the traditional equations of physics are not necessarily oriented and other languages try to offer similar approaches.

Modelica [MOD 11] and AltaRica [ARN 00] (not reduced to its DataFlow fragment) belong to this language category. This expressive power has a counterpart. For AltaRica, calculations remain decidable in all the cases, but the performances attained from the first generations of tools were significantly degraded. In Modelica, it is not generally possible to automate the search of all the failure combinations or all the sequences of limited size leading to a failure condition.

3.5.4. *Injecting faults in nominal models*

Lastly, a certain number of works are trying to develop tools to extend specifications of the nominal behavior and to carry out the safety analyses starting from the extended models. The aim is not only to factorize the efforts of modeling, but also to guarantee consistency between the various system models built during a project. In practice, the kind of injected faults is limited (it is hard to express a deviation of a component output which is a function of some component inputs) and two main cases of performances are encountered.

In the first case, the specification of the nominal behavior is very detailed (for example, cases of Simulink® or SCADE models used to describe control laws). Failure modes are injected inside this initial nominal behaviour i.e. alternative computations of some output values are added. This inserted code is activated when the failure is present. More often, these models are too expressive and, as indicated previously, the current tools do not make it possible to carry out a system analysis in its entirety [AKE 06, HEI 05].

The initial model must be simplified and when it is manual, this process is neither more economic nor safer than directly making models dedicated to safety analysis.

In the second case, the initial model has a limited expressivity. Let us consider, for example, an AADL model [FEI 06] describing the static architecture of a system.

[BOZ 09] proposes an extension of AADL to deal with the system behaviour when faults occur. In this case, the system models can be analyzed with acceptable performances.

3.6. Conclusion

3.6.1. *An approach to assess the safety of systems tested in aeronautics*

Through two cases studies of a different nature (modular integrated avionics, a system with dominating data processing, and an electrical system), this chapter presented how the AltaRica language and associated tools could be used to carry out safety assessments in embedded systems.

The first environments of modeling and analysis came out at the beginning of the 2000s and the approach was found valuable both by research projects and industrial programs. Since 2004, it has contributed, for instance, to the certification of the flight control of Falcon 7X at Dassault Aviation.

We observed that the simplicity of the language makes it available quickly to engineers and that its flexibility makes it possible to carry out failure propagation models of varied granularities, going from the functional view, where only the availability and the loss of functions are considered, to views related to the physics of electric, mechanical, or hydraulic components of the studied system.

Moreover, this simplicity, accompanied by restrictions on the expressivity of the language, is an important factor in being able to assess complex systems with satisfactory response times.

Certain points remain to be improved in the short term, like the modeling of causality loops presented in the section on the electrical system, or the performances of computing times for the largest models.

In the long term, we note that using compositional failure propagation models rather radically changes the way of leading the safety assessment of systems and that this can have repercussions on the overall reliability process.

3.6.2. *Clarification of the system architecture and horizontal exploration of the failure propagation: impacts on the scope of analyses*

Indeed, modeling the propagation of failures consists of characterizing the links between an initial fault, that is not necessarily observable through an internal physical change, and the failures induced, observable through the appearance of deviations compared to the functions that the system should fulfill.

The traditional models (FMEA or fault-trees) were designed to explore this link between causes and effects vertically: inductively, based on the causes in FMEA, or deductively, based on the adverse effects in the fault-trees. This vertical exploration is not constrained: it can follow the physical hierarchical architecture of a system as well as a logical functional decomposition.

Furthermore, these formalisms allow the analyst to be free to increase the granularity of analysis based on the complexity of the system or stage of the achieved assessment process. On the other hand, the explored architecture system, whether it is functional or organic, is not fixed by the model, which leaves the dependencies between components and the perimeter (the set of components) of the explored system implicit. In a complex system, this increases the risk of having information of heterogeneous, incomplete, even occasionally contradictory granularities compared to the perimeter of study.

The approaches of compositional modeling of the propagation of the failures proceed in an horizontal way, between a set of well identified components, which the reliability expert considered to be relevant for the executed stage of analysis. As previously stated, it will be necessary to consider the primary faults of the components and their effects on the functions of the system. But in this case, the transformation of a fault into function deviations in the system imperatively follows the explicit architecture of the model: it is done gradually by transforming a local deviation which is present at the input of each component, into an output deviation.

Thus, the choice of the atomic components of the system and the level of detail of the transformations modeled in each component fix the level of granularity of the analysis once and for all.

This constraint has beneficial effects: it contributes to obtaining more homogeneous, consistent, and probably more complete models in the case of systems including strongly interdependent components. On the other hand, the models of a system are dedicated to a given level of analysis. Changing the level of analysis generally requires changing the model.

This, then, poses the problem of coherence between more or less refined models. This was studied, for example in [KEH 05, BER 09]. [KEH 05] proposes structural patterns that allow the user to abstract elements of architectures satisfying certain properties of fault tolerance. [BER 09] defines a theoretical framework for refining AltaRica models. More recently, tools developed in [LIS 10] allow matching two different models and analyzing whether or not there is a relation of refinement between the combinations of failures generated from each of the two models.

3.6.3. *Clarification of the nominal system characteristics: impacts on the generic definitions of the failure modes*

Two principal options of modeling are offered to the reliability experts when they enter into the details of the transformation models between input/output of each atomic component of the system.

First, they must keep the same approach taken in FMEA or fault-trees. The component propagates only deviations compared to its nominal role in the system, but this nominal role is left implicit. In particular, the barriers to the propagation of the failures are not clarified. This approach in particular is tested in [HEI 01]. Like FMEA or trees, it makes it possible to naturally adapt the level of deviation granularity to the need for analysis and to easily observe the relevant failure conditions.

However, this approach does not necessarily facilitate defining the logic of propagation in each component. This logic depends not only on the current state of the component and the deviations received in input, but also on the nominal functional capacities remaining everywhere else in the system. Thus, it will have, for example, to be deduced from the type of input failure if a prior component raised an alarm. In a similar way, the later components will have to deduce from the output failures of the component if it contains protection mechanisms, which could block the detected failure. This supports neither the local analysis of the propagation, nor the definition of generic components since the models of the components then are very often specialized according to their place in the architecture.

The second option of modeling consists of clarifying not only the rules of failure propagation but also nominal information. The difficulty is then identifying the relevant nominal parameters for study and abstracting them to be able to obtain usable results (i.e. at the same calculable time by the tools and synthetic enough to be readable by the reliability expert).

To achieve this goal, we propose to clarify the normal generic functions of the components employed in the system's architecture and to consider modes of failure for these generic functions. It is this approach that was favored in the models presented in this chapter, and it facilitates the re-use of components as well as the definition of local behaviors.

On the other hand, it requires determining the local generic functions to model. Indeed, each component fills primary functions (for example, "to provide electric power" for a generator) and has side effects (for example "to dissipate heat"). It appears ineffective, even illusory, to seek to model all the generic effects of a component.

In fact, it is enough that the selection of the modeled functions and the granularity of their description make it possible to correctly propagate and observe the system effects to be studied. The way in which the failure conditions can be broken up into a combination of elementary failures of certain system components is thus a first criterion that largely directs the choice of the local functions to be modeled. For example, as the risk "fire" was not studied in our electrical system example, the dissipation of heat was not modeled.

The second criterion concerns having the most complete possible propagation in relation to the need for the analysis. This choice is relatively natural in cases where the perimeter of the studied system includes components with ports of interconnection that are well identified and rather homogeneous from the technological standpoint, i.e. whose primary functions of generation, transmission, or reception treat flows of the same types. This is the case with many " aircraft systems" such as the electrical, hydraulic, or control systems, for which the local potential interactions between components are relatively well-known. Of course, side effects can be incorrectly forgotten or neglected, but the compositional model merits clarifying the nature of the dependencies considered between components where the vertical model remains vague.

In this context, methodological guides are necessary to assist reliability experts in choosing models. Thus, one result of the MISSA project ([MIS 11], [PAP 11]) specifies the alternatives of modeling outlined in this section. [ADE 11] proposes a guide to model systems with a fine enough granularity for multiple physical fields.

3.6.4. *Compositional models of failure propagation: impacts on the overall safety process*

Lastly, the possibility of carrying out and effectively analyzing compositional models of failure propagations integrating nominal information opens the door for many other analyses.

Thus, this type of approach facilitates the study of the impact of potential software errors in system architecture. In the preliminary design phase, this type of hypothetical error is not closely analyzed, using fault-trees for example, because it would lead the expert to manually consider case combinations that would be too large. Detailed software architecture can be more easily depicted by compositional models. The compositional modelling approach can start with the accurate definition of smaller generic components. Then the global model is created by assembling the basic generic components. Finally, the available tools remain applicable to the complex global model and they provide the users with a very rich set of paths of failure propagation leading to failure conditions. [HUM 08] shows how to carry out this type of model and to use it to find critical software requirements for a safe system. Such sets of paths of failure propagation are also used in [BIE 11] to assist the allocation of Development Assurance Level (DAL) to system functions. A tool was developed to deal with the allocation rules which are defined in aeronautics in the ARP 4754 standard [SAE 10].

The proximity of the model of failure propagation to a physical architecture system was also used to study the impact of particular risks on the safety of embedded systems. Indeed, certain particular risks, such as the bursting of a tire or the explosion of an engine, produce debris, which when propelled with a high energy on a given trajectory, can destroy independent components. These trajectories and the lists of destroyed components are calculable with 3D modeling tools, such as Catia. However, if it is really tedious to establish a link between the components of a Catia model and those dispersed or left implicit in a forest of fault-trees, the task is easier when a single well-structured model is considered. The link is then used in the AltaRica model to automatically introduce new common causes of failures and to check that the safety requirements are always maintained [SAG 08].

Lastly, beyond these direct impacts on the process of system safety assessment, the lessons drawn from these various experiments encourage us to explore AltaRica modelings for other needs, such as the analysis of the safety procedures of aircraft operation [MAI 11] or the optimization of the aircraft maintenance [TIA 11], and, if necessary, to develop new tools, in particular to diagnose [GRI 11] or to forecast [TEI 11] the health of an aircraft.

3.7. Special thanks

The authors would like to thank the creators of the AltaRica language without which the work presented in this chapter would not have been carried out: André Arnold, Alain Griffault, Gérald Point, and Antoine Rauzy.

We would also like to thank our colleagues and ONERA students with whom we studied the uses of AltaRica for analyzing the safety of embedded aeronautical systems: Romain Adeline, Eric Bensana, Romain Bernard, Julien Brunel, Charles Castel, Rémi Delmas, Guy Durrieu, Sophie Humbert, Guillaume Infantes, Christophe Kehren, Claire Pagetti, Florent Teichteil, and Laurent Sagaspe. Lastly, we thank our European colleagues from project MISSA and our long-term French industrial partners: Jean-Paul Blanquart, Jean-Marc Bosc, Pierre Darfeuil, Jean Gauthier, Jean-Pierre Heckmann, Sylvain Metge, Marion Morel, François Pouzolz, and Valérie Sartor.

3.8. Bibliography

[ABD 04] ABDULLA P.A., DENEUX J., STÅLMARCK G., ÅGREN H., ÅKERLUND O., "Designing safe, reliable systems using Scade", *ISoLA 2004*, *p.*115-129, 2004.

[ADE 10] ADELINE R., CARDOSO J., DARFEUIL P., HUMBERT S., SEGUIN C., "Vers une méthodologie de modélisation AltaRica pour les systèmes physiques", *Actes du congrès Lambda-Mu'17*, La Rochelle, November 2010.

[ADE 11] ADELINE R., Méthodes pour la validation de modèles formels pour la sûreté de fonctionnement et extension aux problèmes multi-physique, thesis, University of Toulouse, March 2011.

[ÅKE 03] ÅKERLUND O., BIEBER P., BÖDE E., BOUGNOL C., BOZZANO M., BRETSCHNEIDER M., CASTEL C., CAVALLO A., CIFALDI M., CIMATTI A., GRIFFAULT A., KEHREN C., LAWRENCE B., LÜDTKE A., METGE S., PAPADOPOULOS C., PASSARELLO R., PEIKENKAMP T., PERSSON P., SEGUIN C., TROTTA L., VALACCA L., VILLAFIORITA A., ZACCO G., "ESACS: an integrated methodology for design and safety analysis of complex systems", *Proceedings of ESREL*, Balkema, 2003.

[ÅKE 06] ÅKERLUND O., BIEBER P., BÖDE E., BOZZANO M., BRETSCHNEIDER M., CASTEL C., CAVALLO A., CIFALDI M., GAUTHIER J., GRIFFAULT A., LISAGOR O., LÜDTKE A., METGE S., PAPADOPOULOS C., PEIKENKAMP T., SAGASPE L., SEGUIN C., TRIVEDI H., VALACCA-L., "ISAAC, a framework for integrated safety analysis of functional, geometrical and human aspects", *Proceedings of ERTS*, 2006.

[ÅKE 99] ÅKERLUND O., NADJM-TEHRANI S., STAALMARCK G., "Integration of formal methods into system safety and reliability analysis", *17th International System Safety Conference*, August 1999.

[ARN 00] ARNOLD A., GRIFFAULT A., POINT G., RAUZY A., "The AltaRica formalism for describing concurrent systems", *Fundamenta Infor Maticae*, p. 109-124, no. 34, *IOS Press*, 2000.

[BER 07] BERNARD R., AUBERT J.-J., BIEBER P., MERLINI C., METGE S., "Experiments in model-based safety analysis: flight controls", *Proceedings of IFAC Workshop on Dependable Control of Discrete Systems*, Cachan, 2007.

[BER 09] BERNARD R., Analyse de sécurité multi-système, thesis, University of Bordeaux, 2009.

[BIE 04] BIEBER P., BOUGNOL C., CASTEL C., HECKMANN J.P., KEHREN C., METGE S., SEGUIN C., "Safety assessment with AltaRica: lessons learnt based on two aircraft system studies", *Proceedings of World Computer Congress, IFIP*, 2004.

[BIE 11] BIEBER P., DELMAS R., SEGUIN C., "DALculus – Theory and Tool for Development Assurance Level Allocation", *30th International Conference on Computer Safety, Reliability and Security, SAFECOMP 2011*, Naples, Italy, September 2011.

[BOI 06] BOITEAU M., DUTUIT Y., RAUZY A., SIGNORET J.-P., "The AltaRica data-flow language in use: assessment of production availability of a multiStates system", *Reliability Engineering and System Safety*, vol. 91, p. 747-755, Elsevier, Kidlington, 2006.

[BOU 09] BOULANGER J.-L. (ed.), *Sécurisation des architectures informatiques – exemples concrets*, Hermes Lavoisier, 2009.

[BOU 86] BOUISSOU M., "Recherche et quantification automatiques de séquences accidentelles pour un système réparable", *5e Congrès de fiabilité et maintenabilité*, Biarritz, October 1986.

[BOU 91] BOUISSOU M., BOUHADANA H., BANNELIER M., VILLATTE N., "Knowledge modelling and reliability processing: presentation of the FIGARO language and associated tools", *Safecomp '91*, Trondheim, Norway, November 1991.

[BOZ 07] BOZZANO M., VILLAFIORITA A., "The FSAP/NuSMV-SA safety analysis platform", *International Journal on Software Tools for Technology Transfer*, vol. 9, no. 1, p. 5-24, 2007.

[BOZ 09] BOZZANO M., CIMATTI A., KATOEN J.-P., NGUYEN V.Y., NOLL T., ROVERI M., "The COMPASS approach: correctness, modelling and performability of aerospace systems", *Proceedings of SAFECOMP 2009, LNCS 5775*, p. 173-186, Hamburg, 15-18 September 2009.

[DAL 00] DALY D. et al, "Möbius: an extensible tool for performance and dependability modeling", Schaumnurg B.R., Haverkort H., Bohnenkamp C., Smith C.U. (ed.), *11th International Conference, TOOLS 2000, Lecture Notes in Computer Science*, p. 332-336, 2000.

[DUT 97] DUTUIT Y., RAUZY A., "Exact and truncated computations of prime implicants of coherent and non-coherent fault trees within aralia", *Reliability Engineering and System Safety*, no. 58, p.127-144, Elsevier, November 1997.

[FEI 06] FEILER P.H., GLUCH D.P., HUDAK J.J., The architecture analysis & design language (AADL): an introduction, Technical Note, CMU/SEI-2006-TN-011, 2006.

[FEN 94] FENELON P., MCDERMID J.A., NICHOLSON M., PUMFREY D.J., "Towards integrated safety analysis and design", *ACM Computing Reviews*, vol. 2, no. 1, p. 21-32, 1994.

[GRI 98] GRIFFAULT A., LAJEUNESSE S., POINT G., RAUZY A., SIGNORET J.-P., THOMAS P., "The AltaRica language", *International Conference on Safety and Reliability, ESREL '98*, June 1998.

[GRI 06] GRIFFAULT A., POINT G., On the partial translation of Lustre programs into the AltaRica language and vice versa, Research Report No. 1415-06, LaBRI, University of Bordeaux I, November 2006.

[GRI 11] GRIFFAULT A. *et al.*, "AltaRica to analyse relationship between observers and failures model based safety assessment workshop", cisec.enseeiht.fr/images/docs/2011_MBSAW/d2_1100_kuntz.pdf, Toulouse, March 2011.

[HEI 01] HEINER G, MCDERMID J.A., PAPADOPOULOS Y., SASSE R., "Analysis and synthesis of the behaviour of complex programmable electronic systems in conditions of failure", *Reliability Engineering and System Safety*, Elsevier, 2001.

[HEI 05] HEIMDAHL M.P., JOSHI A., WHALEN M., "Model-based safety analysis: final report", *NASA Technical Report*, 2005.

[HUM 08] HUMBERT S., SEGUIN C., CASTEL C., BOSC. JM., "Deriving safety software requirements from an AltaRica system model", *27th International Conference on Computer Safety, Reliability and Security, SAFECOMP* 2008, Newcastle upon Tyne, LNCS 5219, September 2008.

[KEH 04] KEHREN C., BIEBER P., CASTEL C., SEGUIN C., "Vérification par *model-checking* d'un système électrique", *Lambda-Mu*, 2004.

[KEH 05] KEHREN C., Motifs formels d'architectures de systèmes pour la sûreté de fonctionnement, thesis, University of Toulouse, December 2005.

[LIS 10] LISAGOR O., BOZZANO M., BRETSCHNEIDER M., KELLY T.P., "Incremental safety assessment: enabling the comparison of safety analysis results", *Proceedings of ISSC 2010, System Safety Society*, Minneapolis, Minnesota, 30 August-3 September 2010.

[MAI 11] MAITREHENRY S., METGE S., AIT AMEUR Y., BIEBER P., "Towards model-based functional hazard assessment at aircraft level", *International Conference on Safety and Reliability, ESREL'2011*, 2011.

[MCD 94] MCDERMID J., PUMFREY D., "A development of hazard analysis to aid software design", *Proceedings of COMPASS*, 1994.

[McM 93] MCMILLAN K.L., *Symbolic Model Checking*, Kluwer Academic Publishers, 1993.

[MIS 11] MISSA Project, www.missa-fp7.eu, 2011.

[MOD 11] MODELICA, www.modelica.org, 2011.

[PAP 11] PAPADOPOULOS C, SEGUIN C., BIEBER P., BOZZANO M., BOEDE E., BRETSCHNEIDER M., CAVALLO A., DELMAS R., DENEUX J., HECKMANN J.P., LISAGOR O., MOREL M., SAGASPE L., SARTOR V., "Model-based safety assessment for the three stages of refinement of the system development process in ARP4754A", *SAE 2011 AeroTech Congress & Exhibition*, Toulouse, France, October 2011.

[POI 00] POINT G., AltaRica: contribution à l'unification des méthodes formelles et de la sûreté de fonctionnement, thesis, University of Bordeaux, 2000.

[RAU 93] RAUZY A., "New algorithms for fault trees analysis, in reliability engineering and system safety", *Elsevier Applied Science,* vol. 5, no. 59, p. 203-211, 1993.

[RAU 02] RAUZY A., "Modes automata and their compilation into fault trees", *Reliability Engineering and System Safety*, vol. 78, p. 1-12, 2002.

[RTC 92] RTCA, Software Considerations in Airborne Systems and Equipment Certification DO178B, December 1992.

[RTC 00] RTCA, Design Assurance Guidance for Airborne Electronic Hardware DO254, April 2000.

[SAE 10] SAE AEROSPACE, Aerospace recommended practice – Development of civil aircraft and systems ARP 4754a, 2010.

[SAG 05] SAGASPE L., BEL G., BIEBER P., BONIOL F., CASTEL C., "Safe allocation of shared avionics resources", *High Assurance System Engineering (HASE)*, 2005.

[SAG 08] SAGASPE L., Allocation sûre dans les systèmes aéronautiques: modélisation, vérification et génération, thesis, University of Bordeaux, December 2008.

[TIA 11] TIASSOU K., KANOUN K., KAANICHE K.M., SEGUIN C., PAPADOPOULOS C., "Modelling aircraft operational reliability", *30th International Conference on Computer Safety, Reliability and Security, SAFECOMP 2011*, Naples, September 2011.

[TEI 11] TEICHTEIL F., INFANTES G., SEGUIN C., "Lazy forward-chaining methods for probabilistic model-checking", *International Conference on Safety and Reliability, ESREL'2011*, 2011.

[VIN 03] VINCENT A., Conception et réalisation d'un vérificateur de modèles AltaRica, thesis, University of Bordeaux, 2003.

[VIN 04] VINCENT A., GRIFFAULT A., POINT G., "Vérification formelle pour AltaRica, Maîtrise des risques et sûreté de fonctionnement", *lambda-mu 14*, 2004.

Chapter 4

Polyspace®

4.1. Overview[1]

The development of embedded systems encompasses a wide range of practices and development methods. For critical projects intended for extremely robust applications, the production of very high quality software is essential. Development and test teams need to carry out code reviews and test the target operating system.

But is this really enough? What would happen if a critical problem went unnoticed during the development phase in the operational use of the code?

Methods based on mathematical techniques can partly enable us to alleviate doubt.

The application of formal methods based on code verification using *abstract interpretation* can provide development teams with a precise report on the piece of code that will work well, and those that have or risk having errors. This chapter presents the practical application of these techniques for software verification.

In the context of the application of this quality improvement technique, this chapter explores the techniques that can be used with Polyspace®, one of the MathWorks® products.

Chapter written by Patrick MUNIER.

1 This chapter describing the Polyspace® product and its main uses in industry comes from an internal MathWorks report from January 2011.

4.2. Introduction to software quality and verification procedures

Most software development teams aim to maximize both quality and productivity. During software development, however, there are always three interdependent variables: cost, quality and time. Generally, the criticality of the final use determines which of these variables is to be favored. With classical testing processes, development teams usually try and reach their quality objective by testing all the modules of an application until each meets the required quality level. Unfortunately, this procedure is often stopped before the quality objectives are reached as the allocated time and/or budget has expired. Furthermore, code verification and testing that is only carried out at the end of the development process is not enough to attain maximum quality and productivity. It must be integrated into the whole verification process, while respecting time and budgetary constraints.

The verification and validation process (often referred to as V&V) of complex embedded systems consists of determining whether the requirements of the software are correctly and completely implemented, and whether they are traceable to the system requirements.

The main objective of a V&V process is to analyze and test the software during the development and test phase to ensure that the software does indeed perform its intended functions correctly, to make sure that it does not perform any unintended operations, and to provide information about its quality and reliability.

Software V&V also enables us to determine whether the software meets its technical specifications, as well as its needs in terms of security, robustness and reliability. V&V analyses, reviews, demonstrates, or tests everything that is produced by software development. This process is often described in the form of a V diagram, as presented in Figure 4.1.

The specific classes of software defaults that, strictly speaking, can be introduced into the design and coding phases of the V&V procedure are coding and design errors, as well as execution or *run-time* errors, which are software errors that are often considered latent.

Coding errors occur due to improper implementation of code when compared to the requirements or specification. For example, the specifications require the existence of three fault conditions, but the code only implements two of them. Design errors occur when the code does not behave as it should for a specific operating environment. For example, a variable that is a 16-bit type for which the program produces a result greater than the highest value that can be represented on 16 bits.

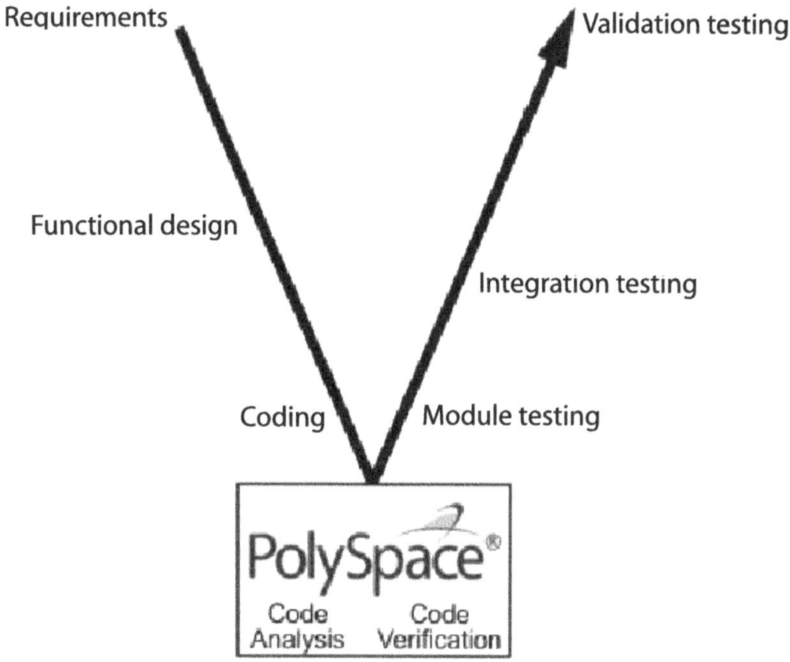

Figure 4.1. *Software verification and validation*

Run-time errors are due to a faulty code, for example, a mathematical operation leading to division by zero. Run-time errors are insidious, as these faults can exist in the code, but are hard to detect. The code therefore seems to be working normally, but this can result in unexpected failures of the system. Software with such errors cannot be considered robust.

Here are a few examples of run-time errors:

– *non-initialized variable*: if a variable is not initialized before it is read, it can contain an unknown value;

– *access outside the bounds of an array*: this represents reading or writing of data outside the bounds of the allocated memory;

– *de-reference of a null pointer*: this occurs when attempting to reference memory with a pointer that is NULL. Any de-reference of this pointer leads to a crash;

– *incorrect arithmetic calculation*: these errors are caused by arithmetic errors such as *overflows*, divisions by zero, or for example by taking the square root of a negative value;

– *concurrent access to shared variable*: this error is caused by two (or more) variables declared in different tasks that want to access the same memory location;

– *illegal type conversions*: these can lead to data corruption;

– *dead code*: even though dead code (code that cannot be executed) does not really represent a run-time error, it may be important to explain its presence. We can note that standards, such as DO-178B [RTC 92] prohibit the presence of non-justified dead code;

– *non-justified infinite loops*: these errors are caused by incorrect guard conditions on program loop operations (for example *for*, *while*, etc.) and may result in system hangs or halts.

The following sections present various techniques that can be used to minimize or eliminate these errors in software, and describe how these techniques can be integrated into a V&V process to create high-quality embedded software.

4.3. Static analysis

Static analysis, or static verification, is a generic term that can be applied to any tool that derives the dynamic properties of a program but without executing it. Most static analysis or verification tools only verify the complexity of the software and look for constructions that can potentially be dangerous.

As described in [BES 10], these tools can usually find errors in the code, but will miss errors that are strongly dependent on data flow (such as run-time errors or design errors).

Due to the fact that they do not exhaustively analyze all possible behaviors, these tools are not considered sound (they miss errors) and thus produce "*false negatives*" (undetected real errors). Decreasing the risk of false negatives will increase the probability of *false positives*, that is to say the production of alarms that are not linked to real problems in the code.

The use of classic static analysis tools can contribute to the automation of part of the verification procedure, but this advantage must be carefully weighed against the ability of tools to generate false negatives and thus not detect errors.

4.4. Dynamic tests

Dynamic tests verify software execution flow. Wagner [WAG 06] describes this methodology and explains the applications of these tests according to the dimension of types (both functional and structural) and granularity, such as unitary tests, integration tests, and functional tests.

Dynamic tests lead to the creation of test cases or test vectors as well as the execution of the software with these tests. A comparison between the results and the standard or expected behavior is carried out. Wagner [WAG 06] also includes a summary of varied compiled statistics with respect to the efficiency of dynamic tests. His analysis shows that the efficiency of dynamic tests is around 47%. In other words, less than half of potential errors, on average, are detected by dynamic tests.

4.5. Abstract interpretation

Abstract interpretation is a formal mathematical method based on techniques that enable us to abstract the semantics of a program. It can be an efficient method for certain types of software verification.

The benefits of abstract interpretation can be illustrated in a simple way by the multiplication of the three big whole numbers in the following equation:

$$4{,}586 \times 34{,}985 \times 2{,}389 = ?$$

Although the calculation of the solution to the problem takes time if it is done by hand, it is possible to apply the rules of the multiplication sign to determine that the result is negative. The determination of the sign of this calculation is an example of the application of abstract interpretation. The technique enables us to verify a set of properties of a final result, such as the sign, with certainty without having to do the whole calculation.

By applying these rules, it is also possible to know that the result of this calculation will never be positive. This property of the program at this point can be useful if, for example, the result of this calculation must be used as a parameter of the function calculating the square root of a number. Similarly, abstract interpretation can be applied to the semantic of software to prove certain properties of the software. Abstract interpretation enables us to relate the techniques for conventional static analysis with dynamic tests by verifying certain dynamic properties of the source code without having to run the program.

Abstract interpretation takes into account all the possible behaviors of a program, that is to say all the combinations of values, and can thus determine whether the program contains certain categories of run-time errors. The results that come from abstract interpretation are considered as sound as their mathematic proof.

4.6. Code verification

Code verification or code verifiers based on abstract interpretation can be used as static analysis tools to detect and mathematically prove the absence of certain run-time errors in the source code, such as overflows, divisions by zero and out-of-bounds array access.

Verification does not require us to run the program, instrument the code, or write test cases. In [COU 02], Patrick Cousot described the success of the abstract interpretation in the analysis of programs. In [DEU 96], Alain Deutsch described the application of this technique to a commercial solution.

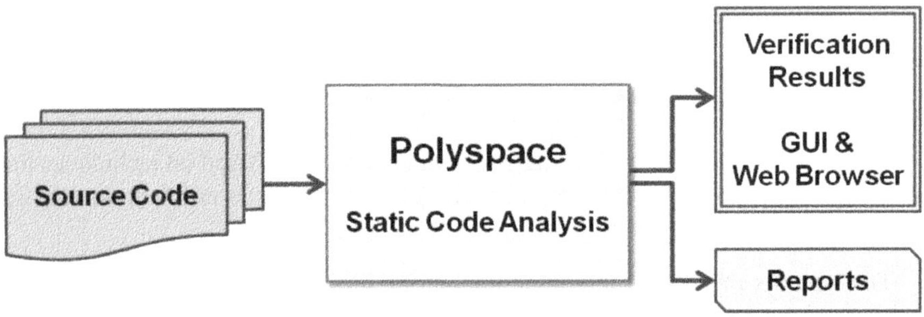

Figure 4.2. *Polyspace® input and output*

To describe the use of the code verification based on abstract interpretation, we will use Polyspace® [MAT 10a and b], from MathWorks® in this chapter. Polyspace® is a code verifier that detects and proves the absence of certain run-time errors, such as overflows, divisions by zero and out-of-bound array access.

As presented in Figure 4.2, Polyspace® accepts input from source code C, C++, or Ada. Polyspace® starts by examining the source code to determine where run-time errors could potentially occur.

Polyspace® then uses a color code to indicate the status of each element in the code (see Figure 4.3).

Color	Explanation
Green	Indicates that the code is reliable – no run-time error can occur
Red	A run-time error exists in this point of the program
Gray	Dead or unreachable code
Orange	A run-time error is possible in this program

Table 4.1. *Color code used by Polyspace®*

Polyspace® results in green mean that the code does not contain certain run-time errors (see Table 4.1). In the event that a certain or possible run-time error has been detected, the code is colored red or orange. Software developers and testers can use this information to fix run-time errors. Polyspace® also calculates the bounds of the variables of code and displays in the graphical interface, as Figure 4.4 shows.

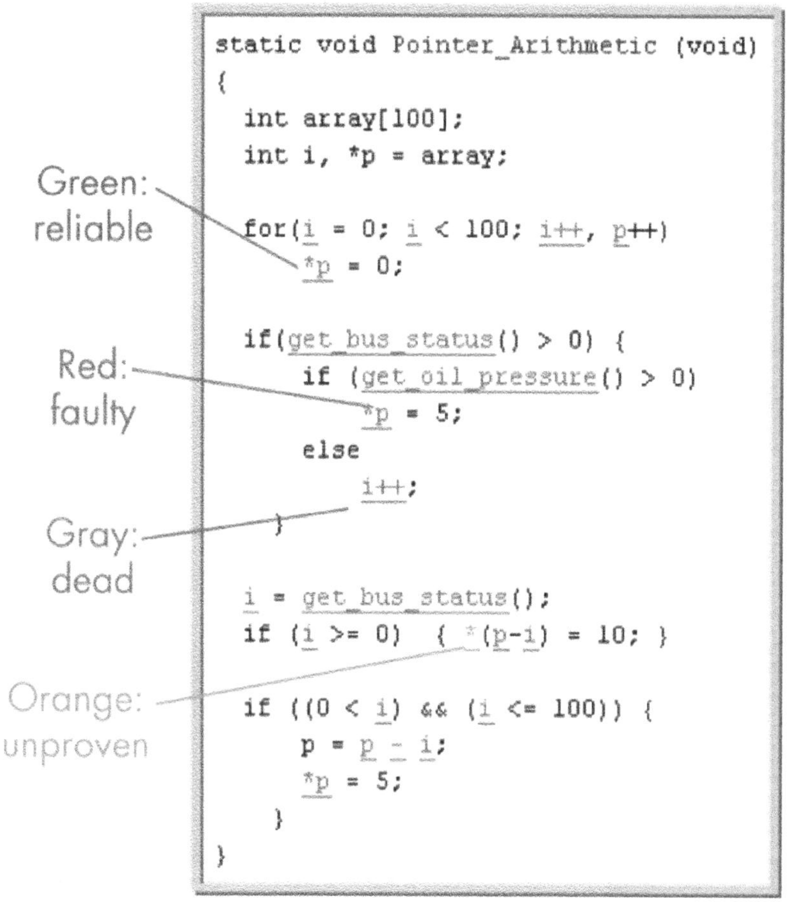

Figure 4.3. *Example of an analysis result provided by the Polyspace® tool (for a color version of this figure see www.iste.co.uk/boulanger/static.zip)*

Polyspace® results can also be displayed in the form of a summary in a dashboard accessible via a Web interface (see Figure 4.5). This functionality is called Polyspace® Metrics.

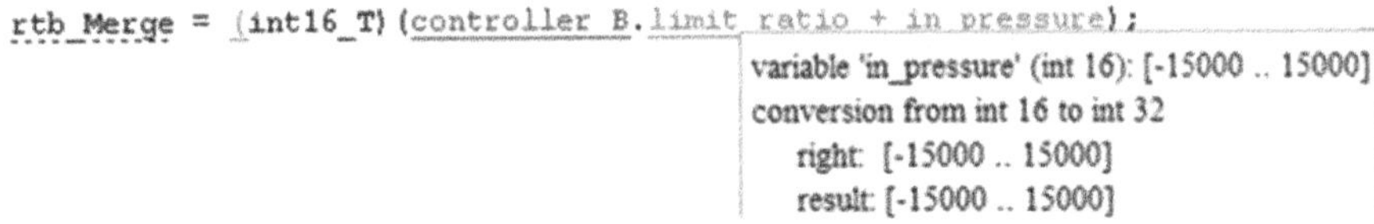

Figure 4.4. *The bounds of the variables displayed in the graphical interface (for a color version of this figure see www.iste.co.uk/boulanger/static.zip)*

Polyspace® can also be used to verify standard coding rules (MISRA-C [MIS 04], MISRA-C++ [MIS 08], and JSF++ [JSF 05]), and produce complexity measures of the code (cyclomatic number of functions, density of commentaries, etc.).

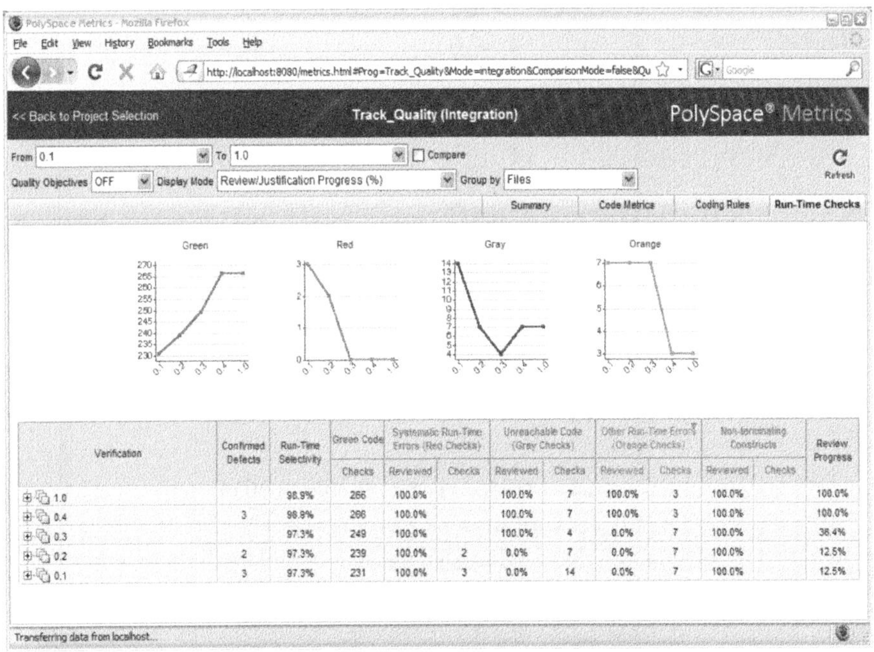

Figure 4.5. *Display of a summary of Polyspace® results in the interface Web Polyspace® Metrics (for a color version of this figure, see www.iste.co.uk/boulanger/static.zip)*

The graphic interface of Polyspace® also includes project management options (see Figure 4.6). This graphic interface enables the user to specify the target environment and compilation settings, as well as other Polyspace® configuration parameters. The tool can also be directly launched on the user's files or via the intermediary of a plug integrated into the Eclipse and Visual Studio development environments.

Polyspace® analyses are often carried out on a dedicated server and users can follow the status of ongoing analyses from the work station. Polyspace® uses the multicore characteristics of machines to accelerate verifications.

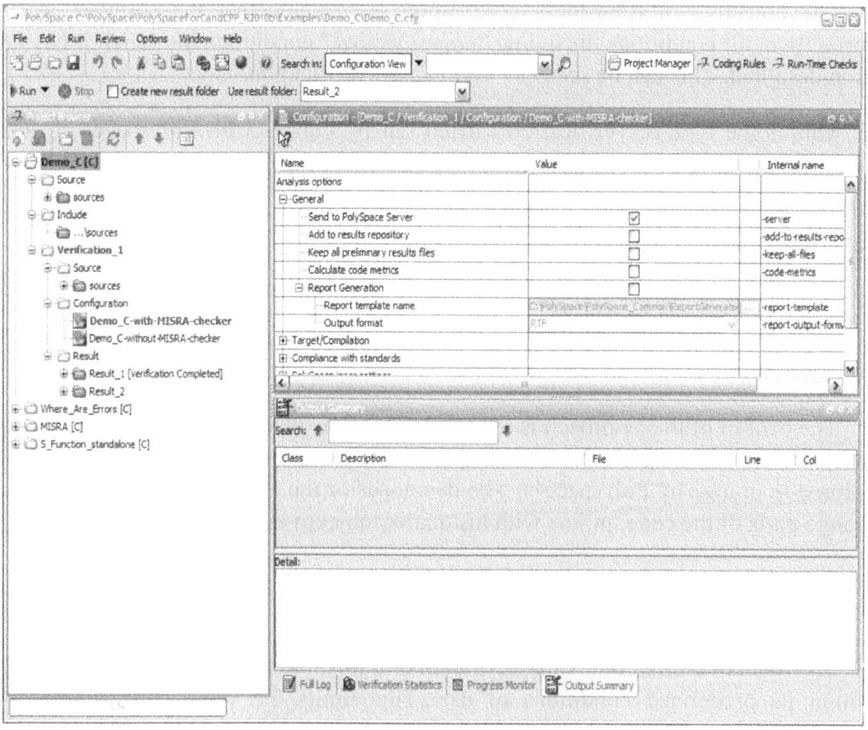

Figure 4.6. *The project view of Polyspace®*

4.7. Robustness verification or contextual verification

Polyspace® users can carry out so-called robustness or contextual verifications. Robustness verification helps to show that the software functions correctly in all of its conditions of use, whereas contextual verification shows that it works correctly in the nominal conditions of its use.

The verification processes can include both types of verification. For example, developers can carry out robustness verifications in their files as early as possible in the development cycle, even while they are writing their code. Later, the team can carry out contextual verifications on larger software components. In addition to this, quality objectives can involve the verification of coding rules and define minimum quality thresholds to be reached.

4.7.1. *Robustness verification*

The objective of robustness verification is to show that the software works correctly in all conditions, including "*non-standard*" conditions for which it has not been designed. This can be seen as "*worst case*" verification.

This is the type of verification carried out by default with Polyspace®. In this case, Polyspace®:

– assumes that the input of functions is not bounded and can take on all of the values permitted by their types;

– initializes global variables with all the values permitted by their types;

– automatically stubs the missing functions by assuming that their output is not bounded and that they can take on all the values permitted by their types.

The benefit of this approach is to make sure of the behavior of the program in all cases. Its disadvantage is that it can produce more non-proven results (therefore colored in orange by Polyspace®). The developer or the tester must then inspect the orange parts of the code, in line with his quality objectives.

4.7.2. *Contextual verification*

The objective of a contextual verification is to show that the software functions within its predefined conditions of use. This limits the verification domain to specific values for certain variables.

The realization of contextual verifications with Polyspace® will reduce the number of unproven results.

Various techniques are made available to the user to help him or her define the verification context of his or her application. The user can:

– use the DRS (*data range specification*) module to specify the ranges of function input and global variables, thereby limiting verification to these cases;

– write a main program by hand to model the call sequence of its functions, rather than base it on the main program automatically created by Polyspace®;

– provide manual stub functions that emulate the behavior of missing functions, rather than base it on stubs that are automatically created by the product.

4.8. Examples of Polyspace® results

The following examples demonstrate the efficiency of *abstract interpretation* with Polyspace®.

4.8.1. *Example of safe code*

Let us consider the function presented in the context of Figure 4.7.

```
int where_are_errors (int input)
{
int x, y, k;

k = input / 100;
x = 2;
y = k + 5;
while (x < 10)
      {
      x++;
      y = y + 3;
      }

if ((3*k + 100) > 43)
      {
      y++;
      x = x / (x - y);
      }

return x;
}
```

Figure 4.7. *Example of safe code*

The objective is to identify run-time errors in the function *where_are_errors ()*. This function carries out various mathematical calculations, contains a loop, *while* and an instruction, if.

We can note that all variables are initialized and used. In line 17, a division by zero can occur if $x = y$.

Given the control structures and mathematical operations on x and y, is it possible that x equals y?

where_are_errors.c

```
1    int where_are_errors(int input)
2    {
3    int x, y, k;
4
5    k = input / 100;
6    x = 2;
7    y = k + 5;
8    while (x < 10)
9            {
10           x++;
11           y = y + 3;
12           }
13
14   if ((3*k + 100) > 43)
15           {
16           y++;
17           x = x / (x - y);
18           }
19
20   return x;
21   }
```

Figure 4.8. *Polyspace® result on safe code (for a color version of this figure, see www.iste.co.uk/boulanger/static.zip)*

As shown in Figure 4.8, Polyspace® has proven that this code does not contain any run-time errors. This is because line 17 is executed only if the condition (3*k + 100 > 43) is true. Knowing that the value of y is dependent on that of k, Polyspace® establishes that at line 17, when x is equal to 10, y will always have a value that is greater than 10. Consequently, there can be no division by zero on this line of code. This result is obtained without having to run the code, write test cases, instrument the code or debug the code. Polyspace® identifies all the places in the code where a run-time error could occur. These places are underlined (see Figure 4.6).

In our example, knowing that no execution error can occur, all of the code is underlined in green. In line 17, for example, the fact that the division operator is underlined in green proves the absence of overflow and division by zero at this point in the program.

4.8.2. *Example: de-referencing of a pointer outside its bounds*

In the example of code in Figure 4.9, the table *ar* has been allocated for 100 elements. The pointer, *p*, points to the first element in table *ar*. The *for* loop increments from 0 to 99 via the index, *i*. In the loop, the pointer, *p*, is also incremented. At the output of loop *for*, the index, *i*, points to the 100th element of table *ar*. The need to store data (value 5) causes a run-time error.

```
void out_of_bounds_pointer(void)
{

int ar[100];
int *p = ar;
int i;

for (i = 0; i < 100; i++)
        {
        *p = 0;
        p++;
        }

*p = 5;

}
```

Figure 4.9. *Example of code with an out-of-bounds pointer*

Figure 4.10 shows the results of the verification of this code with Polyspace®. We can note the instructions underlined in green in lines 8, 10, 11 and part of line 14.

The indications provided by Polyspace® in these places mean that no run-time errors can occur, hence the text is green. Conversely, Polyspace® identified in line 14 that data was written in a memory zone via pointer *p*, and shows that this operation causes a run-time error by underlining it in red.

out_of_bounds_pointer.c

```
1     void out_of_bounds_pointer(void)
2     {
3
4     int ar[100];
5     int *p = ar;
6     int i;
7
8     for (i = 0; i < 100; i++)
9             {
10            *p = 0;
11            p++;
12            }
13
14    *p = 5;
15
16    }
```

Figure 4.10. *Polyspace® results for an out-of-bounds pointer (for a color version of this figure, see www.iste.co.uk/boulanger/static.zip)*

4.8.3. *Example: inter-procedural calls*

In the example in Figure 4.11, *d* is a pointer to an integer incremented with 1 at the beginning of the function *comp()*. It is then used as a denominator of a division to determine the value of the advance variable, to then be recursively passed as a parameter of the same function, *comp()*. Verifying whether the division of line 6 will cause a division by zero requires an inter-procedural analysis to be carried out to determine which value will be passed in the parameter of the function *comp()*.

In the example, function *comp()* is called twice in the *bug_in_recursive* function. In the event where *comp* is called with 10 as the parameter, **d* is a variable incremented from 11 to 49. In this case, there is no division by zero in line 6.

When *comp()* is called with −4 as the parameter, **d* is a variable incremented from −3 to 49. At the fourth call of *comp()*, **d* is worth 0, which causes a division by zero in line 6.

```
1     void comp (int* d)
2     {
3
4     float advance;
5     *d = *d + 1;
6     advance = 1.0/(float)(*d);
7
8     if (*d < 50)
9     comp (d);
10
11    }
12
13    void bug_in_recursive (void)
14    {
15    int x;
16
17    x = 10;
18    comp ( &x );
19
20    x = -4;
21    comp ( &x );
22    }
```

Figure 4.11. *Example of code with interprocedural calls*

A syntaxical verification cannot detect this run-time error. *Abstract interpretation* (see the result given in the context of Figure 4.12) with Polyspace® enables us to prove that this code is exempt from run-time errors except in lines 6 and 21. Division by zero is shown by an orange color in line 6 because the division operation is correct when *comp()* is called in line 18 and incorrect when *comp()* is called in line 21.

This example illustrates the unique ability of *abstract interpretation* to carry out an inter-procedural analysis that takes into account the effects of pointers for distinguishing calls of functions that are problematic from those that are not.

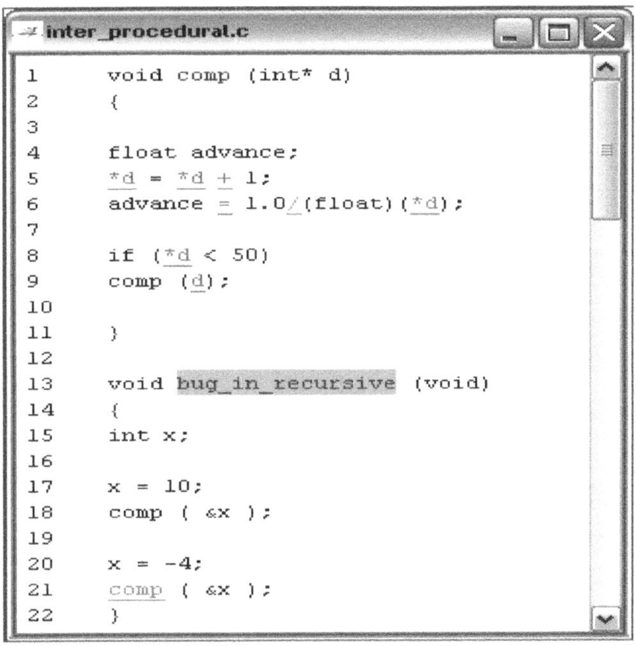

Figure 4.12. *Polyspace® results on the interprocedural analysis of a recursive function*

4.9. Carrying out a code verification with Polyspace®

Code verification with Polyspace® can be carried out in various contexts. They range from individual verification of functions or files, to the verification of more or less luminous components.

The software to be verified may be either handwritten, or produced automatically by a code generator. It could also be a "*mixed code*" of generated and handwritten (manual) codes.

Figure 4.13 presents an example of a Polyspace® result on code automatically generated from the *Real-Time Workshop Embedded Coder* code generator from MathWorks®. The family of Polyspace® products includes the *Model Link* product, which enables us to connect Polyspace® to certain code generators, facilitating its use in *model-based design* environments. The *Model Link* product facilitates both the launching of Polyspace® based on modeling environments and the review of results by enabling a return to models from the Polyspace® results.

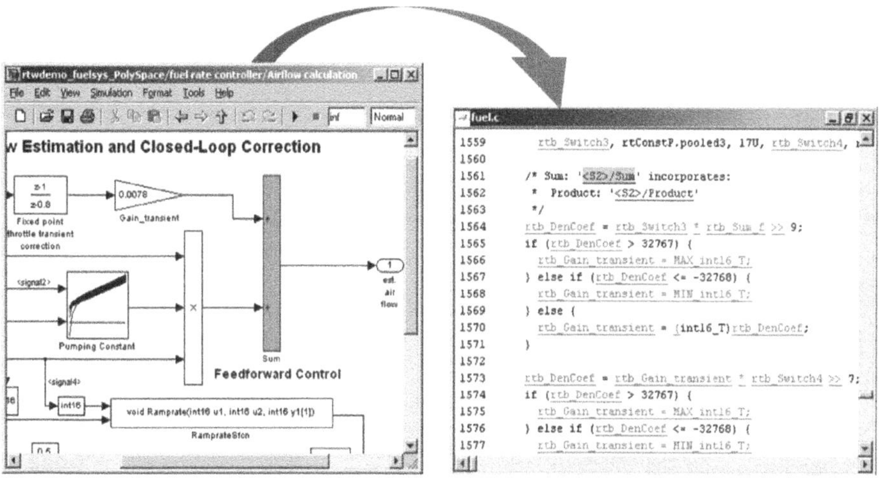

Figure 4.13. *On the left: model of a system in Simulink. On the right: code generated from a code generator and colored by Polyspace®* *(for a color version of this figure, see www.iste.co.uk/boulanger/static.zip)*

With the growing complexity of embedded systems, we note that there is increasingly widespread use of techniques based on models (so-called *model-based design* techniques) with automatic production of code by a code generator. In this approach, control algorithms are mainly modeled and simulated using *block diagrams* and state machines. The code is then automatically generated from these models and deployed onto embedded microprocessors in target machines. The code produced can nonetheless be a mixture of generated code and manual code. In critical systems, the state-of-the-art literature regarding V&V techniques recommends using simulation and verification, both for the model and the code. Verification of the code is of particular importance in mixed codes.

Automatically generated code can contain references to encapsulated manual code (for example for performance purposes). Conversely, generated code can also be integrated into manual code, for example, in an environment of software development. This mixed code must be verified for the code, as opposed to the verification for the model, to ensure the robustness of the system. This is due to the possible presence of a run-time error in the manual code that cannot be detected in the model. A run-time error can propagate itself from the manual code to corrupt the part of the code that was generated automatically.

Other considerations are worth taking into account in mixed code environments:

– errors linked to the use of data from incorrect calibrations;

– signals that do not respect the bounds provided in their specifications;

– transitions between incorrect states;

– memory corruptions;

– mathematical operations that produce divisions by zero or overflows. Mixed codes that contain this type of error cannot be considered robust.

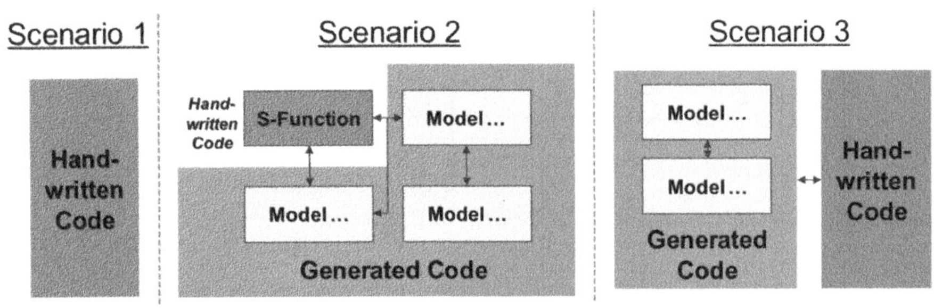

Figure 4.14. *Three scenarios in which Polyspace® can be used*

Figure 4.14 gives examples of scenarios in which Polyspace® can be used. The first is an environment where the whole code is written manually. The second is a mixed code where manually written code is encapsulated in an S-function to be integrated into the model. The third scenario is a mixed code where the automatically generated code is inserted in an environment of software development.

4.10. Use of Polyspace® can improve the quality of embedded software

Modern development and test processes take into account a large variety of development methodologies. The personal preferences and experiences encountered, good or bad, often dictate the processes implemented by the development teams. Tools are often integrated with each other with the help of scripts.

At the heart of these processes, however, we frequently find development and test methods that enable to produce high quality software, ideally free of errors. For example:

– verification of coding standards;

– early verification of software to find issues as soon as possible;

– identification of portions of code which do not contain errors, are all stages which will increase the quality of critical embedded software.

4.10.1. *Begin by establishing models and objectives for software quality*

As the systems they develop are becoming increasingly sophisticated, software development teams must meet rigorous software quality objectives, either dictated by the company itself or imposed by its clients or national regulation authorities.

For these objectives to be reached, and ideally to reach zero defects, it is necessary to define a quality model. This model establishes specific objectives for software quality, with metrics and thresholds to be reached. These objectives provide a mechanism through which teams can communicate their objectives and the state of their development, both internally and with people who are external to their company.

4.10.2. *Example of a software quality model with objectives*

The *Software Quality Objectives for Source Code* [NIS 96] document describes an example of software quality models comprising well-defined objectives. It was written by different automobile constructors and parts manufacturers in partnership with MathWorks®. This quality model includes a list of objectives linked to software quality, which require the use of a tool for the static analysis of code.

Conformity with these objectives is evaluated according to the following criteria:

– a *quality* plan is in place;

– there is a *detailed design*;

– complexity code metrics have been defined and verified;

– adherence to coding rules has been demonstrated (for example, respect of MISRA-C [MIS 04]);

– the code does not include an unattainable branch code;

– the code does not include a construction that does not end (infinite loop, for example);

– systematic run-time errors have been corrected;

– potential run-time errors have been corrected or justified;

– data flow analysis has been carried out.

On the basis of these criteria, the document defines six objectives for software quality (known as *software quality objectives* or SQOs).

SQO-1 only recommends reaching a restricted subset of criteria, whereas SQO-6 requires all criteria to be satisfied. The selection of an adapted SQO level depends on:

– the criticality of the project;

– the quality processes used within the company; these can, for example, require the application of the *capability maturity model integration* (CMMI) or *software process improvement and capability determination* (SPICE);

– whether or not standards such as CEI/IEC 61508 [IEC 98], ISO 26262 [ISO 11], CENELEC EN 50128 [CEN 11], or DO 178B [RTC 92] are needed.

Table 4.2 describes the different criteria applicable to SQO objectives. Certain criteria have associated thresholds, the crossing of which signifies that the SQO objective has been reached.

For example, to reach SQO-1, the table shows that we need to have defined a *quality plan*, written a *detailed design*, and verified code complexity metrics.

To reach SQO-2, it is necessary in addition to this, to respect an initial set of coding rules, to have corrected systematic execution errors, and to have verified that there are no more never-ending constructions in the code.

To reach SQO-3, it is also necessary to show that the code does not comprise unattainable branch codes (*dead code*).

4.10.3. *Use of a subset of languages to satisfy coding rules*

General-purpose languages, such as C and C++, have been designed to develop a large variety of applications, from office automation applications to critical embedded applications.

With extensions such as C99, or extensions linked to compilers such as Visual C++ and GNU, these languages have evolved to support constructions that are increasingly distant from the original language. These have posed difficulties in the verification of complex applications.

The more complex languages are harder to verify, be it by hand or with an automatic tool. To simplify this verification, most standards such as CEI/IEC 61508 [IEC 98], CENELEC EN 50128 [CEN 01], and ISO 26262 [ISO 11, Table A.3] limit the use of a language to a restricted subset.

To meet these standards, a development team must only use language constructions that are authorized by the standard. For example, the quality model illustrated in Table 4.2 recommends certain coding rules of the MISRA-C:2004 [MIS 04] to be adhered to. In addition to making the code easier to verify, coding standards also tend to make the code easier to read, maintain and carry on other targets. Generally, the adoption of a coding standard does not require that all the rules of this standard are verified.

The quality model described above defines two subsets of rules of coding standard MISRA-C:2004 [MIS 04] The first subset, which is required for objectives SQO-1 to SQO-4, includes rules such as:

– *8.11: The static storage class specifier shall be used in definitions and declarations of objects and functions that have internal linkage*;

– *8.12: When an array is declared with external linkage, its size shall be stated explicitly or defined implicitly by initialization*;

– *13.3: Floating-point expressions shall not be tested for equality or inequality*;

– *20.2: Dynamic heap memory allocation shall not be used.*

The second subset, required for levels SQO-5 and SQO-6, includes rules such as:

– *8.7: Objects shall be defined at block scope if they are only accessed from within a single function*;

– *9.2: Braces shall be used to indicate and match the structure in the non-zero initialization of arrays and structures*;

– *13.1: Assignment operators shall not be used in expressions that yield a Boolean value*;

– *20.3: The validity of values passed to library functions shall be checked.*

4.10.4. *Use of Polyspace® to reach software quality objectives*

The elimination of certain types of execution errors is a fundamental part of numerous software quality models. In the SQO model described in Table 4.2, all the potential execution errors to correct or to justify increase as the SQO objectives become more rigorous.

Criteria	*Software quality objectives* (SQO)					
	SQO 1	**SQO 2**	**SQO 3**	**SQO 4**	**SQO 5**	**SQO 6**
A quality plan is in place	X	X	X	X	X	X
There is a detailed design	X	X	X	X	X	X
Code complexity metrics have been defined and verified	X	X	X	X	X	X
Adherence to an initial set of coding rules has been demonstrated	X	X	X	X	X	X
Adherence to a second set of coding rules has been demonstrated					X	X
Systematic execution errors have been corrected		X	X	X	X	X
Code does not include a construction that does not end		X	X	X	X	X
Code does not include an unattainable branch code			X	X	X	X
Potential execution errors of an initial subset have been corrected or justified				X	X	X
Potential execution errors of a second subset have been corrected or justified					X	X
Potential execution errors of a third subset have been corrected or justified						X
Data flow analysis has been carried out						X

Table 4.2. *A software quality model comprising criteria required to reach predefined objectives*

The *Polyspace® Metrics* dashboard, based on a Web interface, provides specific views enabling us to follow quality objectives according to different thresholds defined by the SQO model.

To reach the SQO-2 quality model, the code cannot contain systematic run-time errors or a construction that does not end. To achieve this, the Polyspace® results must not contain a red error.

To further increase software quality and reach the SQO-3 level, the code must not include unreachable branches, which implies that Polyspace® results must not contain gray code.

As the unproven code (code colored in orange by Polyspace®) may not always be a problem, the SQO model establishes different thresholds to define the number of operations of this type that can remain uncorrected and non-justified in the code for objectives SQO-4, SQO-5 and SQO-6.

For example, SQO-4 requires that 80% of potential divisions by zero are proven to be safe or justified. If Polyspace® automatically proves that 70% of these operations are safe (colored in green), then reaching SQO-4 requires an additional manual review to demonstrate that a further 10% of these operations are safe or justified. For the SQO-5 objective, the threshold increases to 90%. For SQO-6, it is 100%. The SQO software quality model authorizes intermediary code deliveries between constructors and suppliers comprising non-proven code. If the final quality objective is SQO-6, then the final delivery must not have divisions by zero that have not been proven safe or justified.

The implementation of a software quality model with well-defined objectives is one of the good practices recommended during the development of critical embedded applications. Polyspace® can be used to reach criteria defined by these models by verifying the conformity of the code to coding rules, by identifying execution errors and branches of dead code, and by enabling teams to quantify potential run-time errors in their applications.

As the complexity of onboard software grows, an increasing number of constructors and parts manufacturers are basing their practices on tools such as Polyspace® for the management of their SQOs. By respecting well-identified *processes* and by applying automatic tools, these companies are able to discriminate between the parts that do contain errors and those that need to be improved.

4.11. Carrying out certification with Polyspace®

Critical software must often be certified. Verification activities must conform to the recommendations of certification standards, such as DO-178B [RTC 92] or

CEI/IEC 61508 [IEC 98]. These standards often mention verification activities such as the verification of coding rules, the detection of execution errors or the detection of design errors.

The improvement of quality, to prove the absence of certain classes of errors, is covered in these standards. For example, this can enable us to reduce or eliminate other verification activities, such as tests at the limits.

4.12. The creation of critical onboard software

Critical software errors, such as run-time errors, can be introduced in the design or coding phases during software development. It is possible that these faults are not detected with traditional dynamic test methods used during phases of unitary tests and integration. Code verification techniques based on *abstract interpretation* provide a tremendous assistance for these points in the development process.

These techniques enable the development teams to know which parts of their code do not have run-time errors, and guide them in identifying the parts of the code that have errors or the risk of errors. The use of these techniques is part of the global V&V phase and allows for an incremental improvement in quality. Their use represents an important step in the context of a software development process that targets zero-default software.

4.13. Concrete uses of Polyspace®

Since its creation at the end of the 1990s, Polyspace® has been used in a large number of projects in varied domains, ranging from aeronautics to defense, medical and industrial machinery to automobiles.

The projects have involved the verification of software written in C, C++ and Ada languages for codes written manually, generated automatically, or a combination of the two. Verification has included unit-by-unit analysis (for example, a verification by C files or C++ classes) to guarantee the robustness of the units verified, or the verification of integrated software components to guarantee their reliability in their call context.

This section presents examples of practical uses of Polyspace® for varied applications. Subsections 4.13.1 to 4.13.3 present the use of Polyspace® in three key industries: the automobile, aerospace and medical industries.

Subsection 4.13.4 provides general examples of use in varied sectors that are concerned with the quality of onboard software.

4.13.1. *Automobile: Cummins engines improve the reliability of their motors' controllers*

Electronic engine controllers are important components in modern motors. However, developing reliable engine controllers that function without problems under varied conditions is a challenge.

The software development and verification techniques used at Cummins Engines were historically based on manual stages, including code reviews and white box tests followed by component and integration tests. Although most errors were detected by the system tests, therefore late in the development process, the complexity of the software made the identification of these errors very difficult, and possibly only at the cost of important *debugging* efforts.

To take into account this complexity, Cummins created an approach in product lines to favor the re-use of software.

For the design of engine control systems, Cummins developers used the *Simulink* and *Stateflow* products from MathWorks®. The automatic generation of code reduced the development time and costs.

For software verification, Cummins engineers decided that it was not enough to focus on component tests and system tests. They needed to increase the efficiency of unitary tests and improve the global process of software tests.

They considered the example of a truck manufacturer who was encountering undue motor stops as a result of a software error in the motor controller. This same constructor had spent three weeks unsuccessfully trying to isolate the problem.

The idea to use Polyspace® as a diagnostic aid was then put forward. Polyspace® detected a problem on an array index that could be decreased in a Stateflow state diagram until reaching a negative value, causing a re-initialization of the controller (see Figure 4.15).

In Figure 4.15, the operation in the Stateflow graph can lead to access to an element of an array outside its bounds:

```
if (status = = error_type_2) {
 error_table[index]+1;
 status_table[index++];
}
```

Cummins also used Polyspace® on other motor control software that was in the development phase, which enabled staff to identify run-time errors early in the development process. In particular, Polyspace® detected data overflow for certain variables of the program. Some of these results were occurring on manual code, but also on automatically generated code when the two types of code were integrated into the complete system.

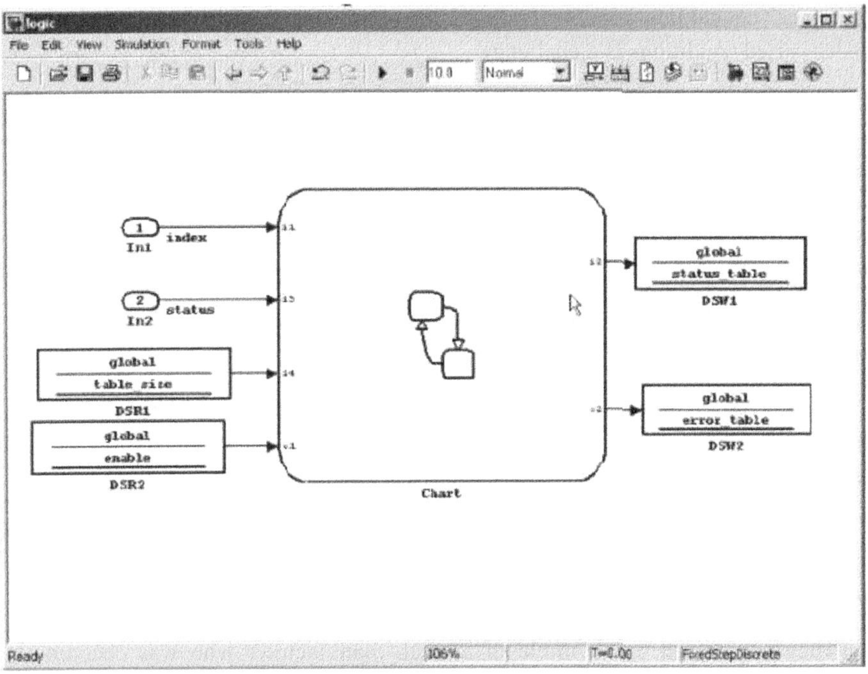

Figure 4.15. *Example of Stateflow design where an index can be decreased until it becomes negative*

4.13.2. *Aerospace: EADS guarantees the reliability of satellite launches*[2]

EADS *Launch Vehicles* is part of EADS *Astrium Space Transportation*, and its mission is to launch satellites into orbit. The reliability of launches is a critical aspect for all space missions.

EADS develops part of the code internally and integrates the code developed by sub-processing companies. If guaranteeing the quality of the code developed internally is difficult, the evaluation of the quality of the code developed by other companies is even harder. To reduce the risk of the presence of run-time errors, EADS used Polyspace® for the Ada programming language to verify a few 100,000

2 See [MAT 10a].

lines of code developed internally and by subprocessing companies. The objective was to verify the totality of the source code and to detect run-time errors as early as possible in the development process. In particular, *overflow* errors, the use of non-initialized data, and conflicts of access to variables shared between several tasks were verified.

Before using Polyspace®, the EADS engineers needed to define, write and execute numerous test cases to run as much code as possible with the aim of detecting errors. With Polyspace®, the code was analyzed exhaustively without having to be run. This enabled engineers to locate operations in the code that systematically produced errors, and others that could lead to problems in certain conditions.

4.13.3. *Medical devices: a code analysis leads to a recall of the device*[3]

The sophistication and complexity of onboard software present in medical devices is increasing. Modern *pacemakers* can have up to 80,000 lines of code on board and infusion pumps more than 170,000 lines. These devices must function with maximum reliability and safety.

The *Federal Drug Administration* (FDA) is the American governmental body tasked with the surveillance of medical devices. It has published several directives and recognized many standards addressing good software development practices. More recently, the FDA established a software laboratory with the *Center for Devices and Radiological Health*, one of the missions of which is to identify software coding errors in devices that could be subject to recall.

One of the investigation techniques retained by the software laboratory is to examine the source code to identify the possible cause(s) of problems in medical devices.

A famous example of a software problem that is feared in medical devices is the one that concerned Therac 25. This device diffused a dose of radiation that was too high for numerous patients. The real-time operating system did not support the notion of messages exchanged between the different *threads* of the program but used global variables instead. Insufficient protection of these global variables caused an erroneous value during the use of the device. The program contained another error that caused the overflow of a whole 8-bit counter. These errors caused the diffusion of radiation doses that were 30 times higher than the prescribed dose.

---

3 See [MAT 10b].

Governmental regulation agencies such as the FDA and certain domains of industry recognize the value of sound verification techniques (that are able to prove the absence of certain errors in software) and use tools based on these principles.

During the *Embedded Software Reliability Workshop* at the 2010 *International Symposium on Software Reliability Engineering* conference, the FDA presented the results of the use of tools such as Polyspace® for the verification of onboard software in medical devices. The title of the presentation was "Applying static analysis to medical device software". The static analysis tools used reported 127 potential problems.

These potential problems were classified in the following way:

– 28 de-references of NULL pointers;

– 36 readings of non-initialized variables;

– 29 illegal transformations of data (*cast*);

– 20 branches of unattainable code (*dead code*).

For the FDA, static analysis showed the lack of good design and development practices for software onboard this device.

The results were integrated into the report submitted to the *Office of Compliance* and led to a recall of the device.

4.13.4. *Other examples of the use of Polyspace®*

Polyspace® has been used by numerous companies in as many industries. The first users were those for whom quality and reliability were important criteria. Here are other examples described briefly.

4.13.4.1. *Tackling sudden acceleration problems*

In 2010, drivers reported to Toyota that their vehicles were prone to sudden accelerations. A company called Exponent carried out an in-depth evaluation of the software driving control of the gas. Polyspace® was used in this process. The preliminary conclusions were reported to the American Governmental Committee for Energy and Commerce in May 2010. In this report, Exponent mentions (14):

> "Polyspace® provides the ability to prove mathematically if a certain class of run-time errors does not exist in the source code and to identify parts of the code where those errors exist or the proof cannot be completed. Polyspace®'s analysis is equivalent to running every possible test scenario, but does it in such a way that no test cases,

instrumentation or real execution of the source code is required. This allows for an exhaustive testing of the software that would otherwise have required several million hours of drive testing to identify.

The Polyspace® analysis was carried out for critical modules, which were identified as controlling the opening angle of the gas command. The analyses carried out have, for the moment, detected no run-time error at the language level".

4.13.4.2. *Use of Polyspace® on Ada code*

NATO's Hawk Management Office is responsible for the improvement programs of the ground-to-air Hawk missile. Polyspace® was used to verify part of the code onboard the missile, which was written in Ada.

In an application that is as complex as a missile, the use of dynamic test cases is not enough to exhaustively test all the values, which could cause a run-time error. The use of these conventional techniques would not have enabled NATO to meet the project time deadline in the search for potential run-time errors. By using Polyspace®, the project team was able to eliminate error-prone constructs in the code that could lead to execution errors.

4.13.4.3. *Ensure the reliability of diesel engines*

Delphi Diesel Systems develops diesel injection systems for all types of vehicles. The major concern in the context of this development is the very high amount of calibration data that need to be incorporated. A typical application of around 200,000 lines of code contains thousands of calibration data. These are constants that can be modified after the fact by calibration teams, which then affect the behavior of the entire application. These characteristics make testing these applications extremely complex.

By using Polyspace®, Delphi Diesel Systems are able to detect run-time errors in software onboard injection systems during the unitary test phase, all the while eliminating robustness tests.

4.13.4.4. *Verification of software in airbag opening systems*

Elesys is developing an airbag suppressor system that detects the weight of the passenger and activates or deactivates the airbag accordingly, to avoid injuring children. In the past, Elesys based itself on code reviews to identify potential problems.

The Elesys engineers then used Polyspace® to carry out an exhaustive analysis of their code. Polyspace® analyzes all functions and operations by taking into account all the possible values for variables. The team is now focusing its reviews on places where Polyspace® detected red and orange operations that under certain conditions systematically lead to a run-time error.

4.13.4.5. *Validation of control– command software in nuclear power stations*

L'Institut de Radioprotection et de Sûreté Nucléaire, the French radioprotection and nuclear safety institute, carries out specialized evaluations of nuclear and radiological risks. The IRSN was formed from the union of the IPSN (protection and nuclear safety institute) with the OPRI (protection against ionizing radiation office).

The IRSN has, in particular, used Polyspace® to exhaustively validate applications for verifying neutrons in 900 megawatt power stations. The use of Polyspace® has enabled it to verify the absence of run-time errors, such as divisions by zero, non-initialized variables, and overflows.

4.13.4.6. *Verification of Nissan motor control software*

Nissan and its parts manufacturers spend a lot of time ensuring the robustness of software during vehicle test phases. In the past, Nissan has encountered software problems linked to the growing complexity of onboard systems.

Nissan now uses Polyspace® on software provided by its parts manufacturers and finds around five problems per project, each project comprising around 30,000 lines of code or 100 K of ROM.

4.14. Conclusion

This chapter gave us the chance to present the Polyspace® tool, discuss its implementation, and give a few examples of industrial applications.

4.15. Bibliography

[BES 10] BESSEY A., BLOCK K., CHELF B., CHOU A., FULTON B., HALLEM S., HENRI-GROS C., KAMSKY A., MCPEAK S., ENGLER D., "A few billion lines of code later: using static analysis to find bugs in the real world", *Communications of the ACM*, vol. 53, no. 2, p. 66-75, 2010.

[CEN 11] CENELEC EN 50128 Railway applications - Communication, signalling and processing systems - Software for railway control and protection systems, 2011.

[COU 02] COUSOT P., "Abstract interpretation: theory and practice", *Model Checking Software*, *Computer Science*, vol. 2318, p. 1-3, 2002.

[DEU 96] DEUTSCH A., *Static Verification of Dynamic Properties*, SIGDA, 1996.

[IEC 98] IEC, IEC 61508: Functional safety of electrical/electronic programmable electronic safety-related systems, International Standard, IEC, 1998.

[ISO 11] ISO, ISO/FDIS26262, Road Vehicles – Functional Safety, ISO, 2011.

[JSF 05] JOINT STRIKE FIGHTER, Air Vehicle C++ Coding Standards for the System Development and Demonstration Program, Document No. 2RDU00001, Rev C, December 2005.

[MAT 10a] MATHWORKS, Polyspace Code Verification Products, MathWorks, 2010 (available at: www.mathworks.com/polyspace).

[MAT 10b] MATHWORKS, Using Polyspace to Implement the "Software Quality Objective for Source Code Quality" Standard, MathWorks, 2010.

[MIS 04] MISRA, MISRA-C:2004, Guidelines for the use of the C language in critical systems, Technical Report, The Motor Industry Software Reliability Association, October 2004.

[MIS 08] MISRA, MISRA-C++:2008, Guidelines for the use of the C++ language in critical systems, Technical Report, The Motor Industry Software Reliability Association, 2008.

[NIS 96] NIST, Reference Information for the Software Verification and Validation Process, National Institute of Standards and Technology, 1996.

[RTC 92] RTCA, DO-178B/ED-12B, Software Considerations in Airborne Systems and Equipment Certification, Version B, RTCA, 1992.

[WAG 06] WAGNER S., *A Literature Survey of the Software Quality Economics of Defect Detection Techniques*, ACM Press, 2006.

Chapter 5

Escher Verification Studio Perfect Developer and Escher C Verifier

5.1. Introduction

Barriers to the widespread adoption of formal methods of software development include the perceived difficulty and expense of using them. Software tools are needed to make it easier to use mathematical methods.

We have two tools for developing mathematically verified software: Perfect Developer (PD) and Escher C Verifier (eCv). PD takes what may be described as a top-down approach, working from a specification, refining it, and generating source code in C++, Java, C#, or Ada. ECv uses a bottom-up approach, being a deep static analyzer for a safe subset of C. Both tools use the Verified Design-by-Contract paradigm to verify each function individually, ensuring that they scale to large software systems.

5.1.1. *Escher Technologies Ltd*

Our organization, Escher Technologies Ltd, was founded by Dr. David Crocker in 1995.

Our aim was to make mathematically verified software development productive. In particular, we wanted to use modern automated reasoning technology to discover

---

Chapter written by Judith CARLTON and David CROCKER.

verification proofs. We needed to make it accessible to ordinary software engineers, that is, to those who knew about engineering and about computing, but who had limited mathematical knowledge.

As software was becoming more and more widely used, and as individual software systems were becoming many times larger than before, we considered it important to create software that would be *correct by its method of construction*, and which, furthermore, *could be proved to be correct*.

5.1.2. *Needs*

Software development tools were clearly needed to facilitate these objectives. From the beginning, we set out to design our tools with a view to overcoming the barriers to the widespread adoption of formal methods. We aimed to produce tools to make mathematical methods relatively easy to use, avoiding mathematical notation as far as possible, and relatively quick to use, saving the developers' time by having an automated theorem prover as opposed to an interactive prover.

Additionally, we knew from experience that one of the more difficult facets of software development by any method (at least, any method in which the developer is not the same person as the end-user) is elicitation of the correct requirements for the software. Our way of dealing with this was to enable a fully functional prototype of the software system to be developed rapidly for approval by end-users, with speedy modification of the prototype, if required.

Following some years of research and development, the Escher Tool was previewed in September 1999 at the World Congress of Formal Methods. This tool was released in March 2001 under the name Perfect Developer. It is now in its fifth major version.

During 2011, Perfect Developer was joined by another tool, Escher C Verifier (abbreviated to eCv). The two products are packaged and distributed as components of Escher Verification Studio, which is available in a commercial edition and in a capacity-limited free edition. Commercial users may license one or both tools.

For clarity, the tools Perfect Developer and Escher C Verifier will be described separately here. The examples were verified using version 5.1 of Escher Verification Studio.

5.2. Perfect Developer – its inspiration and foundations

5.2.1. *Design-by-Contract*

In the late 20th Century, the improvements in computer hardware meant that it was becoming easy to write software that was increasingly large and complex. With this accelerating growth in size, eliminating errors from the software was more and more difficult and time-consuming. Computer scientists such as Hoare, Dijkstra, Meyer and many others sought ways of improving methods of development [HOA 69, DIJ 76, MEY 88].

The term Design-by-Contract was introduced with the Eiffel programming language [MEY 88]. DbC is essentially Hoare logic applied at the level of complete functions or other subprograms. The basic principles are as follows:

– for each class method, a precondition and a post condition are specified;

– whenever a class method is called, the client and the method are bound by a contract. The client guarantees to satisfy the precondition at the point of call; in return the method guarantees to satisfy the post condition at the point of return;

– when class inheritance is used and inherited methods are overridden in derived classes, the relationship between the contracts of the overridden and overriding methods must satisfy the Liskov Substitution principle [LIS 94];

– additional elements such as class invariants and assertions may be included in the specification in order to assist in ensuring that contracts are honoured.

The Design-By-Contract principle is a powerful addition to software development techniques, but it can be hard to ensure that all contracts are honoured. When applied to software developed in traditional programming languages and enhanced by notation for contracts, typically only limited verification of contracts is possible, because:

– the specified contracts are frequently incomplete, often because the notation used is not expressive enough;

– even with class invariants and assertions, insufficient information is available to prove that contracts are honoured;

– contracts are most naturally and simply expressed in terms of a simple abstract data model of a class, whereas the programming language only allows the implementation model to be described;

– the presence of unconstrained polymorphism and reference aliasing introduce too many unknowns;

– it is not certain that honouring the contracts is sufficient to ensure that the program meets its specifications.

5.2.2. *Verified Design-by-Contract*

Verified Design-by-Contract [CRO 04a] as supported by Perfect Developer and the Escher C Verifier, avoids these issues.

While we liked the idea of Design-by-Contract, we thought it would be even better if the contracts specified could be mathematically proved to be honoured.

In Perfect Developer:

– contracts are completely specified, without constraining the implementation unnecessarily;

– contracts are expressed in terms of an abstract data model, which has a specified relationship to the implementation model (if different);

– additional elements such as ghost methods, recursion variants and post-assertions complete the information needed to make automatic verification of contracts possible;

– polymorphism and aliasing occur only on demand, instead of by default;

– required or expected behavior of methods or sequences of method calls may be described, in addition to the basic contracts.

By extending the Design-by-Contract principle in this way, using advanced automated reasoning technology, Perfect Developer provides verification from functional requirements right through to code.

5.2.3. *Perfect Developer*

Perfect Developer is based on the principle of writing formal specifications for systems and developing implementations that are proven mathematically to meet the specifications. The specification language used is called *Perfect*. We do not claim that the *Perfect* language is itself perfect; our claim is that the language can be used to produce code, which perfectly implements a precise specification.

Perfect Developer uses a fully-automatic theorem prover, thus freeing the developer to make good use of time while the proofs are being created. Proving contracts mathematically tends to still more complexity, with the inevitability of enormous proofs [BUN 06]. The proofs are output to file for inspection if required. When the theorem prover fails to find a proof, it produces information to help the developer to identify the possible problem, sometimes including suggestions for missing preconditions or other parts of the specification.

Perfect Developer can generate code automatically from the verified specification. Automatic generation of code allows the developer to view a fully functioning prototype quickly. Once the prototype exists, the developer will probably wish to check with the end-user that requirements were correctly elicited. Any necessary changes are readily made at the specification level, and the verifier re-run. A significant benefit of this approach is that misunderstandings of requirements are exposed earlier in the software development cycle, which reduces cost.

When Perfect Developer is being used to generate final code but the default implementation generated by Perfect Developer is unsuitable, or where Perfect

Developer is unable to generate an implementation, it is the developer's job to supply the tool with sufficient information about how to implement the specification. For example, it is possible that some optimization may appear desirable, and run-time profiling can be used to pinpoint the areas of importance. Changes are made by adding refinements to the specification, not by modifying the generated code (see Figure 5.1).

Perfect Developer is intended for use in a wide range of applications, most importantly in mission-critical real-time systems. Its theorem prover is purpose-built and has been optimized for real industrial problems.

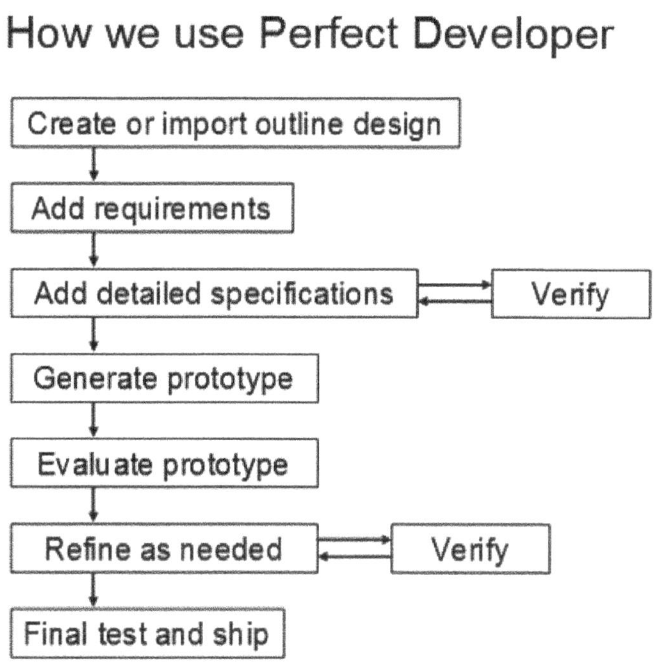

Figure 5.1. *Using PD to generate final code*

5.3. Theoretical foundations

Although Perfect Developer was inspired by Dijkstra's weakest-pre-condition calculus [DIJ 76], it owes more to Floyd-Hoare logic [FLO 67, HOA 69]. Operations in *Perfect* are specified using preconditions and postconditions, and can be composed in sequence or in parallel using the usual rules. In object-oriented designs, polymorphism and dynamic binding are verified by application of the Liskov substitution principle [LIS 94].

5.4. The *Perfect* language

When using Perfect Developer, the developer is required to enter a formal specification of the system or module to be constructed. The specification is written in a formal specification language called *Perfect*.

The design goals of the language included:

– Ease of learning and use. To this end, *Perfect* uses lexical symbols and constructs familiar to software engineers, rather than mathematical notation, and it supports features found in most modern programming languages.

Amongst these features are classes to provide encapsulation, inheritance with polymorphism and dynamic binding, and genericity:

– Expressiveness. *Perfect* includes some constructs from functional programming languages so that, in many cases, specifications can be expressed without recourse to recursion;

– Ease of proof automation. This has to be balanced against expressiveness. For example, full functional programming languages are very expressive, but higher-order reasoning is much more difficult to automate than first-order reasoning. *Perfect* includes only a few constructs (for example, a limited form of left-fold) that require higher-order reasoning.

The following aspects of the system or module to be constructed can be expressed in *Perfect*:

– variables and data structures describing the abstract state: *Perfect* supports familiar data types such as integers, reals, characters, Booleans, and enumeration types. Classes can be declared and class inheritance is supported with polymorphism and dynamic binding if required. Class invariants are supported, as are constrained types (for example, range-limited subtypes of integers and reals);

– functions that retrieve information from the state and from input, defined by preconditions and result definitions;

– operations that modify the state, defined by preconditions, and postconditions;

– expected properties and behavior of the system or module. These are expressed using postassertions, property declarations, and ghost schemas;

– refinement of the abstract state variables into concrete data, where desirable;

– refinement of functions and operations into sequential and/or iterative implementations; Where abstract data has been refined to concrete data, the developer is required to provide refinements for functions and operations that use this data. Elsewhere, refinement is optional. Indeed, it is normally omitted, being carried out only if the developer wishes to use a more efficient algorithm than the one that Perfect Developer would otherwise use.

Although Perfect Developer is intended mainly for specifying software, it is possible in many cases to use *Perfect* to model aspects of the surrounding system as well. This allows properties of the system as a whole to be represented, and hence we can reason about them.

5.5. A Perfect Developer example

We will specify a bounded queue in terms of a sequence. Some expected properties of the specification will be declared and verified. We will then refine the queue by using a ring buffer to store the elements.

5.5.1. *The specification*

The specification (Listings 5.1 and 5.2) declares a class Queue with a single generic type parameter X (which stands for the type of the elements we wish to pass through the queue). The class has an **abstract** section describing the abstract data model, and an **interface** section describing the public interface to the class.

The abstract data comprises:

– an integer variable "capacity" defining the maximum number of elements that the queue is permitted to store;

– a variable "b" of sequence type to store the elements. The sequence type in *Perfect* has the characteristics of both list and array types. For example: like a list, it has variable length and it supports concatenation; but its elements can be accessed by an index just like an array. It also supports insertion or deletion of elements at any position;

– a history invariant, which constrains the variable "capacity" to remain constant after construction. A history invariant is distinguished from a normal invariant by the leading "!" before the keyword **invariant**. It constrains the relationship between the before-and-after values of abstract variables permitted by any operation on the class. The after-values are distinguished by a tick-mark following the variable name;

– a regular invariant, which constrains the length of the sequence to be no more than the declared capacity of the queue.

The symbol "^=" in the listings is read "is defined as".

The interface comprises:

– a Boolean function "empty" that returns true if and only if the queue is empty. In its definition, the unary # operator, when applied to a sequence, returns the length of that sequence. Functions in *Perfect* are always pure, that is, they do not change the state;

– a Boolean function "full" that returns true if and only if the queue is full;

– a schema "add" to add an element. Schemas in *Perfect* are operations that change the state of one or more variables. The "!" symbol before the name of the schema indicates that the current instance of the Queue is modified. In contrast, the parameter "x" is not decorated by a "!" symbol, so it is not modified.

The **pre** clause defines the precondition of the schema. The symbol "∼" is the Boolean negation operator, and "full" refers to the function defined previously. So the precondition states that this schema may only be called if the queue is not full.

The **post** clause defines the postcondition of the schema. In *Perfect*, a postcondition is more than a simple predicate defining the relationship between initial and final states; in particular, it also defines precisely which variables or parts of variables are modified.

```
// Specification of a queue with fixed size limit

final class Queue of X ^=
abstract
    var b: seq of X,            // the queue data
        capacity: int > 0;      // the queue capacity

    !invariant capacity'= capacity;
    invariant #b <= capacity;

interface
    // Test if the queue is empty
    function empty: bool
        ^= #b = 0;

    // Test if the queue is full
    function full: bool
        ^= #b = capacity;

    // Add an element to the end of the queue
    schema !add(x: X)
        pre ~full
        post b! = b.append(x);

    // Remove the element at the head of the queue
    schema !remove(x!: out X)
        pre ~empty
        post x! = b.head,
             b! = b.tail;

    // Build an empty queue
    build{!capacity: int, dummy: X}
        pre capacity > 0
        post b! = seq of X{}
        assert self'.empty;
...
```

Listing 5.1. *Queue specification (part 1)*

The most general form of postcondition is:

change variable – list **satisfy** boolean – expression

However, postconditions of the form **change** v **satisfy** v' = e are used very frequently, so *Perfect* provides the equivalent abbreviated form v! = e.

In this case, the postcondition states that only the value of buffer "b" is changed and that its final value is equal to its initial value with the new element "x" appended:

– a schema "remove" to remove an element from the queue. The "!" decoration before the schema name once again declares that the current instance of Queue is modified, and the "!" decoration on parameter "x" indicates that the parameter is also modified. The **out** keyword declares that the initial value of the parameter is not used, so it does not have to be initialized prior to calling the schema.

The precondition is that the queue is not empty. The postcondition comprises two sub-postconditions, which are separated by a comma. This means that the schema behaves as if the two sub-postconditions (which must modify non-overlapping entities) are satisfied concurrently. These sub-postconditions state, firstly, that the final value of parameter "x" is the value that was initially the first element of the sequence, and, secondly, that the final value of sequence "b" is the tail of the initial value. The *tail* means all the elements of the initial value, except the first:

– A constructor, introduced by the keyword **build**. The leading "!" attached to the "capacity" parameter indicates that the abstract variable of the same name is initialized directly from the parameter, avoiding the need to state this in the postcondition.

The reason for passing a dummy value of type X to the constructor is that this will be needed later, when the queue is implemented as a ring buffer. The precondition is that the capacity is strictly positive, and the postcondition states that the buffer is initialized to an empty sequence. The postassertion (introduced by the **assert** keyword) introduces an additional condition that we expect to hold on the constructed object, and that it should be a necessary consequence of the postcondition.

Also included in the interface section is the specification of some behavior that we wish to verify. Three separate behaviors are specified, as follows:

– The first behavior is expressed using a property declaration, introduced by the keyword **property**. We wish to verify that after an element is added to the queue, the "empty" function always returns false. It has a single parameter "x" whose type is X, the type of elements of the queue. The presence of the parameter means that the property should hold for all possible values of the parameter, subject to any precondition. In this case, the precondition does not refer to the parameter, it merely states that the queue should not be full at the start.

The **assert** clause states the condition we expect to hold. Expression (**self after it**!add(x)) yields the value that would be obtained if we were to take the operand that precedes the **after** keyword (in this case it is **self**, the current instance of Queue) and modify it by executing the operation after the keyword. The pronoun *it* in that operation stands for the expression being modified. So the entire expression after the **assert** keyword means "self after adding an element to it is not empty":

– the second behavior we wish to verify is that if we add an element to an empty queue, the next element we retrieve from the queue will be that element. This behavior could also have been expressed using a property, but as the operations we wish to express modify two different variables, it is simpler to express the behavior using a ghost schema instead.

A ghost schema declaration is just like a regular schema declaration, except that the keyword **ghost** indicates that this is not an operation that needs to be executed at run-time, and therefore no code should be generated for it.

The **then** keyword in the postcondition is used to compose two postconditions sequentially, these being adding the parameter "e" to the queue and then removing an element from the queue into parameter "r". We use an assertion at the end of the ghost schema to express the expected behavior:

– the third behavior we wish to verify is also expressed as a ghost schema, in a similar way.

5.5.2. *Verification conditions generated when the unrefined ring buffer is verified*

Perfect Developer generates 26 verification conditions for this example, and it successfully proves all of them.

```
...
    // Verify that after adding an element, a queue is
    // not empty
    property (x: X)
        pre ~full
        assert ~(self after it!add(x)).empty;

    // If we add an element to an empty queue, the next
    // element we remove should be the one we added
    ghost schema !addToEmptyThenRemove(e: X, r!: out X)
        pre empty
        post !add(e) then !remove(r!)
        assert r'= e;

    // If we add two elements to an empty queue, the
    // first element we remove should be the first one
    // we added
    ghost schema !addTwoToEmptyThenRemove
            (e1, e2: X, r!: out X)
        pre empty,
            capacity >= 2 // we can add 2 elements
        post !add(e1) then !add(e2) then !remove(r!)
        assert r'= e1;
end;
```

Listing 5.2. *Queue specification (part 2)*

Of these, six relate to preservation of the class invariant, and four relate to the validity of the history invariant. Two relate to preconditions of the "head" and "tail" functions being satisfied, and six to the preconditions of the "add" and "remove" schemas being satisfied. One relates to the type constraint "capacity > 0". Three relate to the class invariant of intermediate objects being modified. These are generated when, for example, a schema uses a then-postcondition in which the first part modifies an object in a way that potentially breaks the class invariant, but goes on to call a method in the second part that is entitled to assume the class invariant. One relates to the post-assertion we added to the constructor. The final three relate to successful proof of the three behaviors that we expressed.

5.5.3. *Refining to a ring buffer*

To refine the specification to use a different data representation, we add an **internal** section for the concrete data. See Listing 5.3.

```
// Implement the queue using a ring buffer

final class Queue of X ^=
abstract
    var b: seq of X,              // the queue data
        capacity: int > 0;    // the queue capacity

    !invariant capacity '= capacity;
    invariant #b <= capacity;

internal
    var ring: seq of X,  // the buffer
        hd, tl: nat;     // indices of the first and
                         // next  free elements

    invariant #ring = capacity + 1,
              hd < #ring,
              tl < #ring;

    function b ^=              // retrieve function
        (    [tl >= hd]:  ring.take(tl).drop(hd),
             [tl < hd]:   ring.drop(hd) ++ ring.take(tl)
        );
interface
    ...
```

Listing 5.3. *Queue specification refined to ring buffer implementation (part 1)*

In this case, we declare a fixed-length sequence "ring" (the ring buffer), a variable "hd" giving the index of the oldest element in the queue, and a variable "tl" giving the index of the next free element in the queue. The number of elements in the ring buffer is constrained by the invariant to be exactly one greater then the capacity. Making it one greater means that it is possible to distinguish between an empty and a full queue, but it also means that there will always be at least one unused element. We constrain "hd" and "tl" to be less than the number of elements in the buffer, so that they are always valid indices to buffer elements.

We then redefine abstract variables that we do not want stored. We redefine them in terms of the internal data, using a retrieve function.

In this example, we redefine the original variable length sequence "b" in terms of the internal variables. The construct ([c1]: e1, [d2]: e2 ...) is a conditional expression meaning "if c1 then e1, else if c2 then e2, else...". The expression ring.take(tl) is the usual dot-notation for calling member function "take" of class seq of X, which is the type of "ring". This function returns the first "tl" elements of "ring". Similarly, ring.drop(hd) returns the value obtained by removing the first "hd" elements from "ring". The ++ operator performs sequence concatenation.

Having refined the data, the specifications of the methods now refer to abstract variables that are no longer stored. We have to provide implementations of these specifications. This is done by attaching **via**...**end** statement blocks to the methods concerned. See Listings 5.4 and 5.5. Note that we do not provide implementations for the ghost methods.

```
// Test if the queue is empty
function empty: bool
    ^= #b = 0
    via
        value hd = tl
    end;

// Test if the queue is full
function full: bool
    ^= #b = capacity
    via
        value (tl + 1)%(#ring) = hd
    end;

// Add an element to the end of the queue
schema !add(x: X)
    pre ~full
    post b! = b.append(x)
    via
        ring[tl]! = x,
        tl! = (tl + 1)%(#ring)
    end;

// Remove the element at the head of the queue
schema !remove(x!: out X)
    pre ~empty
    post x! = b.head,
         b! = b.tail
    via
        x! = ring[hd],
        hd! = (hd + 1)%(#ring)
    end;
```

Listing 5.4. *Queue specification refined to ring buffer implementation (part 2)*

To refine function "empty", we merely provide a **value** statement that yields the value hd = tl from the via..end block. We refine function "full" in a similar manner. This time the return value is obtained by incrementing the tail index modulo the length of the buffer and then comparing it to the head pointer.

We refine the "add" schema by storing the new value at ring[tl] and incrementing "tl" modulo the number of elements in "ring". To refine the "remove" schema, we extract the value ring[hd] and store it in the out-parameter "x", then we increment the head index modulo the capacity of the ring buffer.

Finally, we need to refine the constructor (see Listing 5.5). *Perfect* requires that a constructor initializes the entire object being built, which is why we need the dummy parameter. The implementation simply allocates a sequence of the required capacity with all elements initialized to the dummy value, and initializes the head and tail indices to zero.

```
    // Build an empty queue
    build{!capacity: nat, dummy: X}
        pre capacity ~= 0
        post b! = seq of X{}
        via
            ring! = seq of X{dummy}.rep(capacity + 1),
            hd! = 0,
            tl! = 0
        end
        assert self'.empty;

    // Verify that after adding an element, a queue is
    // not empty
    property (x: X)
        pre ~full
        assert ~(self after it!add(x)).empty;

    // If we add an element to an empty queue, the next
    // element we remove should be the one we added

    ghost schema !addToEmptyThenRemove(e: X, r!: out X)
        pre empty
        post !add(e) then !remove(r!)
        assert r' = e;

    // If we add two elements to an empty queue, the
    // first element we remove should be the first one
    // we added
    ghost schema !addTwoToEmptyThenRemove
            (e1, e2: X, r!: out X)
        pre empty,
            capacity > 2 // so we can add 2 elements
        post !add(e1) then !add(e2) then !remove(r!)
        assert r' = e1;
end;
```

Listing 5.5. *Queue specification refined to ring buffer implementation (part 3)*

5.5.4. *Verification conditions generated when the refined ring buffer is verified*

For the refined example, Perfect Developer generates sixty verification conditions, and it proves them all. Twenty-six of them are the same as for the unrefined example, above.

A further nine verification conditions relate to preconditions of the indexing operator [ ], the modulo operator %, and the "take" and "drop" functions. The constructor and the "add" and "remove" schemas are now required to establish and preserve the three internal class invariants as well as the abstract class invariant, resulting in another nine verification conditions.

For each of the refined functions "empty" and "full", a "Returned value satisfies specification" condition is generated. Similarly, for the refined constructor and "add" and "remove" schemas, five "Specification satisfied at end of implementation" conditions are generated (one for each comma-separated postcondition). Additionally, for each schema, a condition "Only variables modified in specification are modified by implementation" is generated.

The variables "hd" and "tl" were declared with type "nat", which is a subtype of "int" containing only non-negative values. This leads to five additional "Type constraint satisfied" verification conditions, one for each assignment to one of these variables.

Finally, the conditional expression used to define the retrieve function did not include an else-part (which would be indicated by an empty guard), so there is a verification condition that at least one guard must be true.

5.6. Escher C verifier

While Perfect Developer starts from the concept of generating code by refining a formal specification, many developers of embedded systems distrust code generation, preferring to write code by hand, most commonly in C. It was to meet their needs that we produced Escher C Verifier, which performs formal verification of C programs.

Our goals were to:

- make it easy to use (keep the annotation simple, provide suggestions);
- handle a sufficient subset of C for SIL2 and higher software;
- make the subset compatible with MISRA-C 2004;
- support a wide range of embedded processors and C and C++ compilers.

The Escher C Verifier extends the C language with additional keywords and constructs needed to express specifications and to strengthen the type system of C.

These annotations are invisible to ordinary C compilers, so the annotated programs can be compiled as normal.

The language subset has been carefully chosen to avoid classic vulnerabilities of the C language [CRO 09a]. The subset includes most constructs permitted by MISRA-C 2004, so developers already coding to this standard find that their code is well on the way to compatibility with the Escher C Verifier code subset. Optional support is provided for some C'99 and a few C++'98 constructs.

Escher C Verifier uses the same infrastructure as Perfect Developer. We therefore chose to retain the Verified Design-by-Contract paradigm [CRO 04a] that we use when verifying *Perfect*. This paradigm requires as a minimum that all function preconditions must be explicitly stated. Although this may seem rather cumbersome for the programmer, the rigorous documentation of preconditions greatly enhances maintainability and re-usability, even when formal methods are not otherwise used. In a critical system, we maintain that when a programmer writes a function call, there should be no uncertainty in his/her mind as to what conditions must be satisfied for the called function to be relied upon to behave as expected.

5.7. The C subset supported by eCv

In the C language, it is easy to write constructs that have undefined behavior or that have implementation-defined behavior [CRO 09a]. It follows that C is far from an ideal choice for writing critical applications. However, the C language definition documents [ANS 89, ISO 99] state fairly clearly the situations in which behavior is not well-defined. This makes it possible to define subsets of C with well-defined semantics. One attempt at such a subset is MISRA C [MIS 04], which has gained widespread acceptance in industry.

Although many of the restrictions imposed by such subsets can be enforced statically, others depend on the inter-relation between different parts of the program, in ways that can only be resolved using mathematical proof.

Whereas some of the MISRA rules are based on notions of "good programming practice" rather than on the avoidance of undefined behavior, we have selected our subset of C solely with verification in mind. Not surprisingly, almost all the constructs that are prohibited in our subset are also prohibited in the MISRA subset although the reverse is not true.

5.8. The annotation language of eCv

The usual way of augmenting programs with formal specifications is to express them as specially formatted comments. In most programming languages, no other choice is available, assuming the requirement that the source text is to be compiled by a standard, unmodified compiler.

However, formatting specifications as comments regrettably suggests to developers that they are unimportant to the functioning of the program. We prefer specifications to have a more central role in the language. Also, most text editors are unable to provide syntax highlighting for specifications expressed in comments.

Fortunately, the C language offers us an alternative: to define specification constructs using macros. In C, macros can be defined so as to expand to nothing when the program is compiled, so that the specification is invisible to a standard compiler but visible to our verifier.

For example, we declare function preconditions using the construct **pre**(*expression*), having made the macro definition:

```
#define pre(x) /* nothing */
```

visible to the compiler.

One small disadvantage of using macros is that, under C89, the number of arguments must be fixed for any particular macro. If the user tries to declare two comma-separated preconditions in a single **pre** clause like this:

```
pre(p1, p2)
```

then the compiler will refuse to compile the program, because it is expecting a single macro argument but finds two. However, the following alternative forms:

```
pre(p1) pre(p2)
pre(p1; p2)
pre((p1, p2))
pre(p1 && p2)
```

all avoid this problem.

5.8.1. *Applying verified design-by-contract to C*

Any attempt to define contracts for functions written in C must address the particular challenges that this language presents. Some of these are discussed here. We use the term *ghost* to refer to an entity that can be used in a specification construct but not in the program text visible to the compiler.

5.8.2. *Arrays and pointers*

Possibly the biggest flaw in the C language is its lack of distinction between a pointer to a single storage location and an array (which, in C, is just a pointer to a block of contiguous locations). A related problem is that when an array is passed as a parameter to a function, the called function has no way of determining the number of addressable elements in the array, unless this information is passed in an additional parameter. However, when writing specifications, it is frequently necessary to refer to the length of an array.

We work around this problem as follows. First, we distinguish between pointers to single locations and pointers to arrays. This we do by using the keyword **array** when declaring an array pointer. Thus, in the following declarations:

int ∗p;
int ∗ array pa;

the variable *p* is a pointer to a single element, whereas the variable *pa* is a pointer to an array. Our tool enforces the correct use of this annotation, for example by prohibiting the use of pointer arithmetic and indexing on plain pointers.

Second, we treat an array pointer as if it is a structure comprising three variables: two ghost integer variables named *lwb* and *upb*, which represent the lower and upper bounds of the array, and the storage pointer itself. The main use of the ghost variables is to describe the function preconditions that ensure array accesses are in bounds.

For example, the precondition of the array access *pa[i] is pa.lwb <= i && i <= pa.upb*. As a further convenience, we provide an additional ghost member *lim* (short for limit) defined as *upb+1*.

A further consideration is that, whereas C allows the value of any pointer to be zero (i.e. pointing to nothing), in many contexts a zero value is inappropriate, because any attempt to de-reference it will lead to undefined behavior. Therefore, we require the programmer to state explicitly that particular variables and parameters of pointer type are allowed to take null values. This is done by decorating the declarations with the keyword **null**. In the absence of such decoration, zero-valued pointers are prohibited.

For example, in the following declarations:

int ∗ array **null** pna;
int ∗p;

variable *pna* may point to an array or be zero, while variable *p* is a non-zero pointer to a simple variable or single element. This approach removes the need to pepper the program with preconditions of the form `p != 0`. Furthermore, when a pointer is de-referenced, we do not need to generate a verification condition that it is non-null unless the user has declared that null is a permitted value for it.

Next, the correct operation of many functions (including library functions like *strcpy*) requires that two or more array pointers passed into the function refer to non-overlapping storage. We allow this to be expressed by augmenting our representation of pointers and array pointers with a ghost member function *disjoint*, which takes another pointer as a parameter and returns true if and only if the result of any legal indexing of the pointers cannot refer to the same element. Of course, C has no notion of functions that are structure members, but we felt here that it was sensible to borrow a small amount of syntax from C++.

Finally, a string literal in C has three different meanings depending on whether it is used as an array initializer, the operand of `sizeof`, or in a context where a value of type `char*` (actually `const char * array` in our subset) is required. We handle this by assigning the *Perfect* type *seq of char* to string literals, but allowing automatic type conversion of string literals to `const char * array` where required. A further refinement allows for the fact that compilers are permitted to perform string pooling, so that identical string literals appearing in different places in the source code are not necessarily disjoint when converted to array pointers.

5.8.3. *Unions*

Although coding guidelines for critical systems often deprecate the use of unions without discriminants, such as those supported in C, we feel that in the absence of alternative facilities (such as discriminated unions, or inheritance and polymorphism), unions are too useful to be excluded altogether. They should be used in a controlled fashion.

In our semantic model, each variable of union type remembers the name of the union member through which its current value was assigned. When a member of a union variable is read, it is a precondition that the member being read must match the remembered name. The remembered name can be queried using the ghost expression `v holds member` where *v* is an expression of union type and *member* is the name of one of the members of its type.

5.8.4. *Side effects*

In writing specifications, it is important to use expressions that are free from side-effects. Unfortunately, side-effects are common in C expressions. We use the term *pure* to indicate that an expression is side-effect free. All expressions used in specifications are required to be pure.

5.8.5. *Quantifiers*

The expression sub-language of C does not provide quantification over types or collections. However, quantification is frequently needed in specifications. We therefore imported quantified expressions from *Perfect* with syntactic modifications to fit better with the style of C.

We allow quantification over types and over collections such as arrays and sets. We use the expression `a..b` to denote the set of integers from *a* to *b* inclusive (or the empty set if $b < a$). For example, the expression:

forall i **in** a.lwb..a.upb :- a[i] > 0

conveys the notion that all members of the array *a* are positive. These expressions can, of course, be used only within specifications.

5.9. Escher C verifier examples

5.9.1. *First escher C verifier example*

5.9.1.1. *Presentation*

Our first Escher C Verifier example is a classic binary search algorithm. We use eCv to show that, subject to the preconditions being met, the algorithm terminates and returns the correct result for all possible inputs.

When writing modular C programs, it is usual to write a header file that defines the interface to the module, and a separate code file that provides the code and data. When a module wishes to call into a second module, the code for the first module imports the header file for the second module using a **#include** directive, so as to access its interface definition. Accordingly, the eCv specifications for public functions are placed in the header file so that, when verifying another module that imports the header file, the specifications of called functions defined in other modules are available.

The header file for this example is shown in Listing 5.6. It is standard C'90 apart from the addition of the pre() and post() annotations. As is usual in C, when a function parameter is a pointer to an array, we have to pass an additional parameter to supply the size of the array. The first precondition describes this relationship. The second precondition requires that the table elements be in strictly increasing order. This could be simplified somewhat by using the built-in function "isndec" instead of using quantifiers. The third precondition is necessary to avoid overflow in the code when we calculate the mid point between high and low array indices. Expression maxof(size_t) is an eCv construct that returns the maximum value representable by type size_t.

```
// File bsearch.h

#ifndef bsearch_h_included
#define bsearch_h_included

#include "stddef.h"        // for size_t
typedef double elem_t;     // defines the element type

// Search a table of ordered elements of type elem_t
// for the closest match above or equal to 'key'
size_t bsearch(const elem_t table[],
         size_t numElems,
         elem_t key)
  pre(table.lim >= numElems)
  pre(forall i in 0..(numElems - 1);
       j in 0..(numElems - 1) :-
        i < j => table[i] < table[j])
  pre(numElems <= maxof(size_t)/2)
  post(result <= numElems)
  post(result == 0 || table[result - 1] < key)
  post(result == numElems || key <= table[result]);

#endif
```

Listing 5.6. *Binary search function header file*

The postconditions define the value returned by the function. The first one says that it does not exceed the number of elements, so that it is either a valid index into the table, or, it is one greater than the last valid index. The second says that if the return value is non-zero, then the table element at position one below the return value is strictly less than the key. The final postcondition says that if the return value is not one-past-the-end of the table, then the table entry at that position is greater than or equal to the key.

The code for this example is shown in Listing 5.7. This is again standard C, apart from the eCv annotations. Each **keep** macro supplies one or more loop invariants, using a semicolon as a delimiter between invariants. The first **keep** macro indicates that the "low" and "high" indices are the right way round and in range, while the other two indicate that the key does not belong in the region of the table below "low" nor in the region above "high". The **decrease** macro provides a loop variant. In this case, the loop invariant is a simple integer expression that must decrease between iterations, but must remain non-negative. These properties, when verified, provide proof that the loop always terminates.

5.9.1.2. *Verification conditions for the binary search example*

Escher C Verifier generates and proves thirty-four verification conditions for this example.

```
// File bsearch.c

#include "ecv.h"
#include "bsearch.h"

size_t bsearch(const elem_t table[],
               size_t numElems,
               elem_t key)
{
  size_t low = 0u, high = numElems;
  while (high > low)
  keep(low <= high; high <= numElems)
  keep(low == 0u || table[low - 1u] < key)
  keep(high == numElems || key <= table[high])
  decrease(high - low)
  {
    size_t mid = (low + high)/2u;
    if (key <= table[mid])
    {
      high = mid;
    }
    else
    {
      low = mid + 1u;
    }
  }
  return low;
}
```

Listing 5.7. *Binary search code*

There are seven instances of array indexing in the specification and code. Each of these gives rise to two verification conditions to ensure that the index is in range. Each occurrence of the division operator gives rise to a condition that the divisor is non-zero.

The for-loop has four invariants, and for each of these, both a "Loop initialization establishes invariant" condition and a "Loop body preserves invariant" condition are generated. For the loop variant, there is one verification condition to ensure that it is non-negative when the loop is first entered, another to ensure that it decreases between iterations, and a third to ensure that it remains non-negative.

There are two instances of integer addition. Each gives rise to two verification conditions, that the result is within the range that can be represented by the type. Finally, there are three "Postcondition satisfied when function returns" verification conditions, one for each postcondition.

5.9.2. *Second example*

5.9.2.1. *Presentation*

Our second example for Escher C Verifier is a reworking of the ring buffer that we expressed in *Perfect* earlier in this chapter.

The example used refinement to show that the ring buffer was equivalent to a bounded sequence to which we added elements at the end and removed elements from the beginning. Unlike *Perfect*, the C language does not naturally support refinement, so it is not surprising that this is less easy to express in annotated C. Nevertheless, it can be done.

In the header file (see Listings 5.8 and 5.9), we first declare a macro constant CAPACITY to represent the maximum number of elements that the queue can hold, and a typedef-name "elem_t" to represent the element type. It is necessary to use a fixed element type because C does not support generics. Next, we declare a type "bufindex_t" for the head and tail indices using eCv's constrained typedef syntax. This behaves like a normal typedef declaration, except for the **invariant** clause that imposes a constraint on the allowed values of this type.

```
// File RingBuffer.h

#define CAPACITY  (20u)    // capacity of the queue

typedef int elem_t;

typedef unsigned int
  invariant (value <= CAPACITY) bufindex_t;

#ifdef __ECV__
extern bufindex_t hd, tl;
extern elem_t ring[];
#endif

// Ghost function to represent the abstract data
ghost( _ecv_seq<elem_t> data(); )

// Test if the queue is empty
bool empty()
  returns(data().count == 0);

// Test if the queue is full
bool full()
  returns(data().count == CAPACITY);

...
```

Listing 5.8. *Ring buffer header file (part 1)*

The functions "add", "remove", and "init" modify static variables "hd", "tl", and "ring". This has to be expressed in the specification, so it is necessary to make these variables accessible in the header file by declaring them **extern**. These declarations are made conditional by means of #ifdef ... #endif so that they are available to eCv but not to the compiler, to avoid breaking the modularity of the program.

```
...

// Add an element to the end of the queue
void add(elem_t x)
  writes(tl; ring)
  pre(!full())
  post(data() == old(data()).append(x));

// Remove the element at the head of the queue
elem_t remove()
  writes(hd)
  pre(!empty())
  post(result == old(data().head()))
  post(data() == old(data().tail()));

// Initialize the queue to be empty
void init()
  writes(hd; tl)
  post(empty());
```

Listing 5.9. *Ring buffer header file (part 2)*

To represent the abstract data and its relationship to the concrete data, we declare a retrieve function "data()". By enclosing the declaration in the **ghost()** macro, we not only make it invisible to the ordinary C++ compilers but also signify to eCv that it is for use in specifications only. The return type is given as "_ecv_seq<**int**>", which is the name by which the *Perfect* type **seq of int** can be accessed in C.

Having defined the abstract data, we can now specify the member functions "empty", "full", "add", "remove", and "init" in terms of this abstract data. The specifications are similar to those used in the *Perfect* version of this example. The "add" function modifies static variables "tl" and "ring", and eCv requires that all such modifications are declared in the specification. This is the purpose of the **writes**-clauses.

In the *Perfect* specification, we used a trailing quote character to signify the final value of the variable concerned in a postcondition. ECv uses the opposite convention, namely that postconditions refer to final values of variables unless the expression concerned is enclosed in the **old** macro.

The "remove" and "init" functions are specified in a similar way.

The code file for this example is shown in Listings 5.10, 5.11 and 5.12. After declaring the head and tail index variables and the array in which elements are stored, the retrieve function for the abstract data is defined in exactly the same way as in the *Perfect* example, but using C syntax. Next come the definitions of functions "empty", "full", "add", "remove", and "init" using standard C syntax. Finally, we have translations to C of the property and of the two ghost schemas that express behavior

```
#include "ecv.h"
#include "Queue.h"

bufindex_t hd, tl;          // head and tail indices
elem_t ring[CAPACITY + 1u];    // the ring buffer itself

// This function returns the abstract data that the
// ring buffer represents
ghost(
  _ecv_seq<elem_t> data()
  returns(  (tl >= hd)
          ? ring.take(tl).drop(hd)
          : ring.drop(hd).concat(ring.take(tl))
         );
)

// Test if the queue is empty
bool empty()
{
  return hd == tl;
}

// Test if the queue is full
bool full()
{
  return (tl + 1u) % (CAPACITY + 1u) == hd;
}

...
```

Listing 5.10. *Ring buffer code (part 1)*

```
...

// Add an element to the end of the queue
void add(elem_t x)
{
  ring[tl] = x,
  tl = (tl + 1u) % (CAPACITY + 1u);
}

// Remove the element at the head of the queue
elem_t remove()
{
  elem_t r = ring[hd];
  hd = (hd + 1u) % (CAPACITY + 1u);
  return r;
}

// Initialize the queue to be empty
void init()
{
  hd = 0u;
  tl = 0u;
}

...
```

Listing 5.11. *Ring buffer code (part 2)*

```
#ifdef __ECV__

// Verify that after adding an element,
// a queue is not empty
void test1(elem_t x)
writes(tl; ring)
pre(!full())
{
  add(x);
  assert(!empty());
}

// Verify that if we add an element to an empty queue,
// the next one we remove will be the one we added
void addToEmptyThenRemove(elem_t e)
writes(hd; tl; ring)
pre(empty())
{
  add(e);
  elem_t r = remove();
  assert(r == e);
}

// Verify that if we add 2 elements to an empty queue,
// the first we remove will be the first one we added
void addTwoToEmptyThenRemove(elem_t e1, elem_t e2)
writes(hd; tl; ring)
pre(empty())
{
  add(e1);
  add(e2);
  elem_t r = remove();
  assert(r == e1);
}

#endif
```

Listing 5.12. *Ring buffer code (part 3)*

that we wish to verify. All three are expressed as functions, and we use the **assert** macro to force eCv to generate the required verification conditions.

5.9.2.2. *Verification conditions generated for the C version of the Queue example*

Escher C Verifier generates and proves 74 verification conditions for this example, nearly 25% more than for the *Perfect* version.

Twelve of them are checks for arithmetic overflow when the integer "+" operator is used. Thirteen are preconditions for the built-in operators [ ] and % and the built-in functions head, tail, take, and drop. A further six relate to preconditions of "add" and "remove" in the three behavior verification functions. Thirty-four relate to satisfying type constraints; for example, when "hd" or "tl" is assigned. Two assert that the values returned by the implementation of "empty" and "remove" satisfy the corresponding **returns**-specification and four assert that the postconditions of "add", "remove", and "init" are satisfied by their respective implementations. The final three are the assertions that we added to describe expected behavior.

5.10. The theorem prover used by Perfect Developer and Escher C verifier

5.10.1. *Logic*

The underlying logic in which verification conditions are expressed is a many-sorted equational logic with partial functions. The *Perfect* language is designed so that most of the verification conditions generated are first-order. This allowed us to base the prover on a traditional first-order theorem prover design using resolution and paramodulation, coupled with a term rewriting system.

Some transformations that are typically applied to proof terms by automatic theorem provers are unsound in the presence of partial functions, even when it is known that the term as a whole is definitely either true or false. Care must be taken to avoid making unsafe inferences in these situations.

A few constructs – such as polymorphism, dynamic binding, and the limited form of left-fold supported by Perfect Developer and Escher C Verifier – are not readily expressible in first-order logic. The theorem prover uses additional lemmas and inference rules to handle the terms to which these constructs give rise.

5.10.2. *Term rewriter*

The term rewriter uses a set of hard-coded rules for actions such as literal evaluation and de Morgan's laws. It also uses a larger set of parameterized rewrite rules, ordered so as to reduce confluence issues. Additional rewrite rules are generated from terms in the prover. Around 60% of verification conditions are proved by the term rewriter alone.

We chose to integrate the proof engine with the rest of the tool, rather than to develop it as a separate entity. The disadvantage of this approach is that it is not straightforward to substitute another theorem prover. However, there are few theorem provers that correctly handle partial functions.

Integrating the theorem prover has a number of advantages. For example, the specification and/or implementation being verified is accessible to the prover, which can inspect it for additional information and/or guidance in selecting appropriate heuristics. When the prover fails to detect a verification condition, it can use the context in which it was generated to formulate a suggestion for amending the specification.

The entire proof engine runs without user intervention, which enables the developers to make maximum use of their time. Perfect Developer was designed from the outset for commercial use, and in practice, most software engineers have neither the skill nor the time needed to develop proofs interactively.

After the run, the theorem prover reports the verification conditions that have been proved. Where it fails to prove a verification condition, it provides information on where the failure occurred; sometimes, it also provides a hint as to how the specification might be modified.

5.11. Real-world applications of Perfect Developer and Escher C verifier

5.11.1. *Safety-critical systems*

5.11.1.1. *General Dynamics (UK) Ltd*

General Dynamics (UK) Ltd has used Perfect Developer to specify and model the software for a safety-critical airborne stores management system. Their need was to meet the requirements of UK defense standard 00-55 to Safety Integrity Level 4 [CRO 09b]. Due to the defense-related nature of the project, we are unable to provide additional detail.

5.11.1.2. *Crowcon Detection Instruments Ltd*

Crowcon Detection Instruments Ltd designs and manufactures gas sensing equipment. Their mission statement is "to be relied on worldwide for the protection of people and plant against gas hazards" [CRO 12].

The sensor units are built around a detector that uses infrared light of a wavelength that is absorbed by the gas of interest. The transmissivity of the air being sampled is compared with the transmissivity of a reference cell. As part of their process to achieve IEC-61508 accreditation for a product, Crowcon needed to be certain that the critical data path (which was between raw readings from the detector and the output produced by the unit) was free from errors. This path includes the following subsystems:

- a linearization table accessed using binary search;
- an averaging filter;
- a conversion function that takes account of calibration data.

Escher C Verifier was applied to this problem. It verified that the binary search function returned the correct value for indexing into the linearization table, the filter function behaved as expected, and (with one exception) the entire data path was free from arithmetic overflow and similar errors. ECv was unable to prove just two verification conditions that show absence of overflow at one point in the averaging filter, and these were verified by manual review instead.

5.11.2. *Teaching university classes*

Universities often teach software development using Java and modeling using UML. Perfect Developer fits well with such courses, since it has an object-oriented basis, it can generate code in Java, and it can import UML models. Because it uses a fully automatic theorem prover, detailed knowledge of proof techniques is not required, so that Perfect Developer is accessible to students with limited mathematical knowledge.

By generating ready-to-compile code, Perfect Developer provides a quick route from writing the specification to running the program. This gives students fast feedback on their work, helping to maintain student motivation.

5.11.2.1. *Griffith college, dublin*

Tony Mullins of Griffith College, Dublin, told us that Perfect Developer greatly improves the quality of the work that students do over the duration of their course. He said [MUL 12] *"Motivated students like it because they can put it on their machines, use an editor to type specifications, and then use the tools to prove the specification correct. They are always surprised that they have missed things in their specifications and it catches them out! After much scratching of the head, they invariably exclaim: Now I see!"*

In summary, we understand from our academic users in general, and from one comparison project in particular [FEI 08], that the following features of Perfect Developer help to make formal methods of software development easier to teach:

– students can run real programs that they have developed themselves;

– automatic theorem prover takes the labor out of verification;

– specification notation is based on programming language constructs;

– only one notation to learn - no need to specify in one and code in another;

– choice of Windows or Linux platform;

– supports an object-oriented approach to software development.

5.11.3. *IT system development: Precision Design Technology Ltd*

We are greatly indebted to John Warren for his help with these pages [WAR 12].

Precision Design Technology Ltd (Precision Design) is a software consultancy company, which agrees with us about the importance of correct specification and the value of formal reasoning. Many organizations have now compiled reports analyzing the success and failure of software and of information system projects. A common theme in all these reports is the importance of a "good" specification: important through its presence in a successful project and important through its absence in a failing one.

We work in close collaboration with Precision Design Technology although, in practice, their approach differs somewhat from ours. With our product Perfect Developer, our approach is iterative and relies on a proven code generator: typically, we specify the application (in incremental parts), generate the code, and demonstrate application correctness by proof and from the running code.

The Precision Design approach, while iterative, does not rely on code generation: they specify the application (also in incremental parts), prove that the specification is consistent, and demonstrate application correctness directly from the specification by reasoning. They call this process *"animation"* and it involves deducing the behavior that the specification would exhibit given any defined sequence of events. They can use each of these "defined sequences" later for user acceptance testing, with the advantage that they then know precisely what the result of each such test should be.

Precision Design consider this animation as being exactly equivalent to testing a provably consistent specification. As with all testing, they state that the animation process shows the presence of errors *but not their absence*. However, this testing is done at the earliest possible point in the project, so mistakes in analysis and specification are easy to correct and require least rework; also, the user view of the specification, as a model of the intended system, is easy for end-users to understand.

5.11.3.1. *Description of Precision Design's approach*

In their approach, called SPECIFY4IT, a specification comprises a hierarchy of four related collections of information, related by rules; the approach subdivides each collection into a number of subsidiary sets of specification data. The approach relates (by rule) the data items in each of these sets to the data in other sets, both in the containing collection and in pre-requisite collections (that is, those that are more senior in the hierarchy). They represent the data in each of the sets using as a formal notation clauses in first-order predicate calculus with some higher-order enhancements.

The Precision Design Technology reasoning process relies on axiomatic proof: there are five axioms that support the reasoning system. They specify the reasoning system, which is an information system in its own right, using its own notation, and build it according to its own rules. It can run as an animation of its own self-specification, though for execution speed they normally replace many of the underlying proof requirements by lemmas. In both their approach and ours, any discrepant application behavior is adjusted exclusively by correcting the application specification.

5.11.3.2. *Language issues*

Precision Design reduces the task of learning a new specification language by providing dialogs through which analysts can interact with each aspect of the system. They translate the information thus provided into the corresponding formal notation on which the reasoning systems will operate. This allows them to demonstrate the consistency of (all or part of) a specification before they undertake any further reasoning, such as animation. This approach conceals all mathematics and reasoning from all end-users. They can animate any (partial or complete) specification that is consistent.

5.11.3.3. *Advantages of a Perfect translator*

Since both our approach and the Precision Design approach are formal and similar in architecture, it is possible to translate from Precision Design's internal notation into the *Perfect* language that we use. Precision Design wrote such a translator, allowing them to transfer swiftly from their approach to ours, thereby gaining the advantages of both.

Precision Design Technology believes that their approach speeds production of a formal specification, and that they can then prove this for consistency *and* animate it to demonstrate correctness.

Then, by translating this proven and correct specification into *Perfect*, they can take advantage of our separate proof engine to confirm that, additionally, the specification satisfies the principles of "Verified Design-by-Contract", as defined by a separate and independent set of verification conditions. They can then use our proven code generators to construct high-quality code directly from the translated specification.

5.11.3.4. *A library model*

Precision Design Technology often uses a model of a library information system to illustrate their approach. This is partly because such systems are well known and hence require no description; and partly because the system, while small enough to illustrate on one sheet of paper, is complex enough to be a realistic example of many aspects of the approach. They also use this as a tutorial model to illustrate the principles of successful addition of modules to enhance an initial system.

The diagram (Figure 5.2) shows part of the tutorial model of the library, in which books, members, and loans can each be added, modified, or deleted; some other functions and rules are also present.

The "soft box" icons show the processes (or activities) that comprise the system. The inter-connecting arrows represent business rules and show dataflow. The system is "event-driven": nothing will happen until a recognizable event in the real world initiates some activity by delivering data to a process in the system. Precision Design identifies those processes that receive data from external events by a short inbound arrow incident on the left edge of the icon. Upon completion, any process may initiate other (connected) processes internally if business rules exist that permit this; this will continue until no further processing is possible.

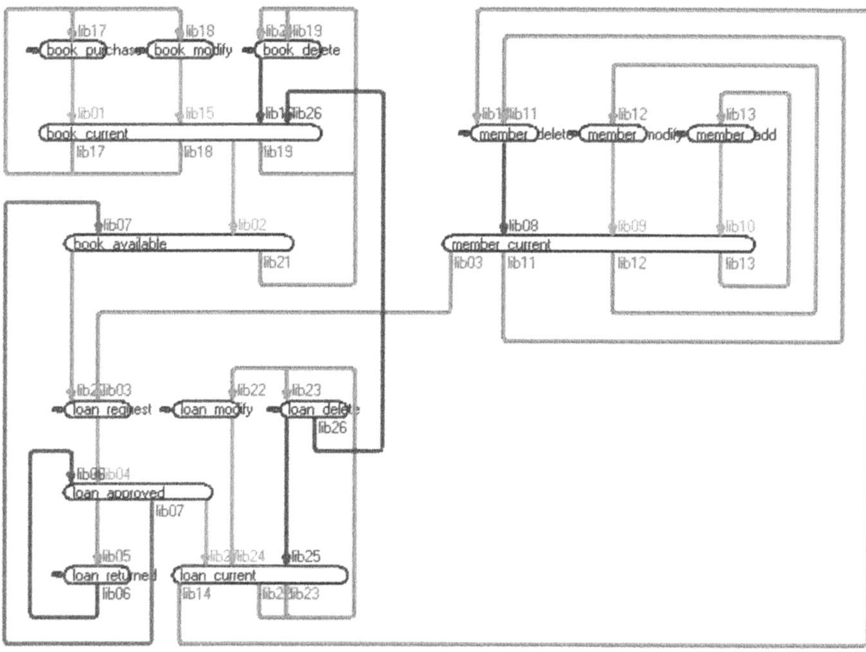

Figure 5.2. *A partial model of a lending library*

The present version of this "tutorial" model expressed in Precision Design's (formal) notation contains 441 "facts" (predicate calculus clauses) and translates to 5778 lines of *Perfect*. This represents an expansion factor of about 13:1.

The translation process produces two separate groups of files: one group that is common to all applications and one group that is specific to the particular application concerned.

Usually, the application-specific group is larger. In this case, there are 24 classes and 4,410 lines of *Perfect*. There are 1,115 verification conditions (some deduced by the translation process and some by the prover itself). All of these are fully provable (100% success) by the automated prover in about 19 seconds on a laptop computer of reasonable speed (1.66 GHz). This is an average of 17 milliseconds per proof or about 58.7 proofs per second. These proofs are usually relatively easy; the longest proof requires only about 0.3 seconds.

In contrast, the fixed or application-independent group is usually smaller, with 13 classes and 1,368 lines of *Perfect*. There are 395 verification conditions deducible. Of these, 388 are fully provable (98.22%) in 889 seconds (slightly less than 15 minutes) on the same machine. This is an average of 2.25 seconds/proof. The longest successful proof requires 12.6 seconds. It is clear that these proofs are much harder than the present "application" proofs, possibly by a factor of 2.25/0.017 or 132:1. However, readers should note that at present, we have not attempted to resolve these seven (395 – 388 = 7) proof problems; no conditions are refutable and, in practice, the software runs without error.

For the model as a whole, the proof success rate is 99.53.

5.11.3.5. *Verification conditions*

There is an important issue connected with the automated deduction of verification conditions from the specification. The successful proof of these conditions means nothing in relation to the correctness of the application in satisfying the end-users' requirements. This is due to the assumption that the specification is correct; any other specification would support deduction and successful proof unless there was some inconsistency in the deduction and proof process. The satisfactory proof of conditions deduced by rule therefore shows consistency (not without value) but not correctness, which is generally more important.

The verification conditions that are most valuable are those provided by the end-users. They define the system properties (verification conditions) on the basis of their expertise and knowledge of the application domain, rather than on the basis of the specification. They, therefore, provide independent corroboration of the specification. Where their proposed properties are provable, this confirms specification correctness in respect of that particular property. Where their properties are refutable, this identifies a demonstrable discrepancy between specification and end-user intention. Where the properties are neither provable nor refutable, there is a clear justification for deeper investigation.

5.11.3.6. *Invariant properties*

We can identify universally valid properties within many application systems. We refer to these properties as invariants and they are actually an abbreviation for a (possibly large) number of separate verification conditions. Invariants also represent user-provided verification conditions (it is not possible to deduce invariants by rule). They represent valuable application domain knowledge that the specification should embody.

For example, borrowing a book from a library does not alter the total library stock; it merely reduces the stock that is available for other borrowers. Verification conditions that require the library stock to remain unaltered despite loans and subsequent returns are (if satisfied) more valuable in showing correctness than verification conditions deduced by rule. These conditions are usually more difficult to formulate (requiring application expertise) and harder to prove. If added to the library model, the invariants relating book stock to book loan and book return activities each require several seconds for proof.

5.11.3.7. *Code generation*

Precision Design's initial concern was to improve the specification process, by ensuring that specifications were provably consistent and demonstrably correct. They later realized that code generation from such specifications was feasible and wrote a separate (non-provable) translator for Java to demonstrate this. The commercial benefits of completely automated and error-free code production within minutes of the production of an accepted specification are obvious. As stated earlier, they then wrote a translator for *Perfect*; though again this is (at present) non-provable. This allows them to translate SPECIFY4IT specifications to *Perfect* and to use the Perfect Developer theorem prover and code generators.

The "library" model, already described, contains 441 "facts" in its specification and translates to 5,778 lines of *Perfect*. If translated to Java source code using Perfect Developer, the result is 13,349 lines of Java, yielding an executable jar file of 413kb. Perfect Developer achieves this in a few seconds.

5.11.3.8. *Subjective comments from precision design*

Precision Design Technology normally tries to ensure that Perfect Developer proves *all* verification conditions successfully. Their reasons for this are that if the verification condition lies within the capability of the prover, then it is foolish to build a system that contains refutable conditions, since these represent known errors. If the condition is not within the prover's capability, then it is appropriate to subdivide the proof into simpler components, each of which is provable, and to use these subsidiary proofs as lemmas to assist the larger or more complex proof. However, Perfect Developer permits the generation of an executable system, directly from the specification, without reference to the embedded proof technology.

Some modules may contain many large and complex verification conditions and hence require considerable proof effort. One, possibly limited, approach is to use faster machines for the proving process. Another approach is to limit the amount of proving undertaken: not so much in the number of conditions stated for proof because this clearly increases proof coverage, but rather in restricting the amount of proving done after each change to the specification. Since Perfect Developer supports the proof of single classes in isolation, there is no need to prove the entire system after any single change to one class. It is enough to repeat the proof of the amended class specification.

Precision Design believes that automated proving is a highly effective strategy for developing high-quality software and information systems. One reason for this is that proof is a rigorous reasoning approach that encompasses all the relevant parts of an information system and is not dependent upon variables containing specific values. In contrast, only testing exercises that specific part of the system under test and the test results may depend critically upon variable values. Proof is therefore generally wider in scope or coverage than comparable testing, and when proof is automated, the procedure will not be only very much faster but also free from human error.

Which human could undertake 1,100+ tests on 4,410 lines of specification (not code) in 19 seconds?

5.12. Future work

We have been continually enhancing the theorem prover, informed by customer experiences and by advances in automated reasoning technology, and we anticipate that this is likely to continue.

Beyond that, and within our over-arching aim of enabling better-quality software, our future work is largely customer-driven. Subject to market demand, we are considering the following.

5.12.1. *Composite projects*

Composite projects - modeling a system in Perfect Developer and implementing it in Escher C Verifier - will be made easier to handle. It will be possible to carry specifications automatically from the model to the implementation.

5.12.2. *Code generation*

Generation of code by Perfect Developer in a subset of C suitable for embedded systems is desirable, but the C language is far from ideal for writing high-integrity code. However, we believe that a small and carefully-chosen subset of C++ could offer substantial advantages, such as encapsulation. We propose to define such a subset, supporting it in eCv, and adding it as a target language for code generation in Perfect Developer. The potential for a standardized treatment of concurrency has emerged with the 2011 C++ standard.

5.12.3. *Concurrency*

We would like to add support for concurrency. At its present stage of evolution, Perfect Developer does not check for deadlock, livelock, or race conditions. It is very hard to reason about concurrency using mutable shared variables. The twin approaches of communicating objects (such as Communicating Sequential Processes) and transactional memory may offer a way forward.

5.12.4. *Extension*

5.12.4.1. *C++*

These language features facilitate better encapsulation and greater type-safety. Some loop invariants will be derived automatically [STA 99].

5.12.4.2. *MISRA*

Reporting MISRA rule violations will be extended such that eCv can serve as a MISRA compliance checker as well as a program verifier.

5.12.4.3. *SPARK Ada*

Perfect Developer can already generate partial code in the SPARK-annotated subset of the Ada language [BAR 03]. We would like to improve the translation of complex specifications to SPARK code annotations. We hope that when support for Ada generics is added to the SPARK language, we will be able to generate complete SPARK Ada code in Perfect Developer.

5.12.5. *Conclusion*

In this chapter, we have given simple code examples of both our major tools, along with some explanation. We have given instances of the use of our tools in industrial systems where safety is considered critical, to a greater or lesser extent (DefStan 00-55 SIL 4 and IEC-61508 SIL 2). We have given a detailed instance of a software consultancy firm, which believes whole-heartedly in the value of formal methods of software development in non-critical applications. Lastly, we have mentioned the advantages of Perfect Developer as a teaching tool in the classroom. We hope that this range of applications gives a fair representation of the capabilities of our software development tools at this stage in their evolution.

Our aim continues to be to raise the quality and reliability of software products to a level where even minor software defects are an exceptional occurrence. In time, we hope that provably-correct software development will become standard practice for most applications in mainstream industry.

By producing tools to reduce the cost of developing critical software, we assist the evolution of software engineering toward delivering software that is at least as reliable as the hardware on which it runs. In theory, software can be constructed to conform perfectly to precisely stated requirements; but it is in the nature of things that hardware must always have some possibility of failure.

5.13. Glossary

The reference manuals are available online [ECV 11, PDL 11] if further information is required.

Assertion

An assertion is a declaration of some condition that we are asserting that must hold ("is true") at that point in an implementation. Perfect Developer will verify this.

Ghost

A ghost (whatever) is one (whatever) which has *no executable code*.

Perfect allows ghost methods, ghost operators, ghost selectors, and ghost schemas.

For instance, when defining functions to represent abstract properties, it sometimes happens that the property is only used in declaring pre-conditions, post-conditions, class invariants, etc.

Postassertion

A postassertion is an assertion given at the end of a method (i.e. after the result or postcondition) that expresses one or more conditions that should be true when the method completes. Each condition should have something to say about the return value (for a function or operator) or the final values of modified objects (for a schema).

For example: say we have a class Money in which we declare a division-by-integer operator that is supposed to round up the result. If our specification really does round up, we expect that as long we start with a non-zero amount, we end up with a non-zero result. We can express this expectation in a postassertion like this:

```
operator /(other: nat): Money
  pre other ~= 0
  ^= Money{(amount + other - 1) / other}
  assert amount ~= 0 ==> result.amount ~= 0;
```

A postassertion is different from a postcondition. A postassertion specifies everything that the client can expect to be true when a method finishes, even in the presence of polymorphism, when different objects may be responding differently to the same message.

Postcondition

A postcondition says precisely which variables have changed, and what conditions are satisfied by the final values of those variables. Postconditions must be complete in every way.

Precondition

A precondition must be met before a method is called.

Property

Part of an object's state, which can be changed.

Schema

A schema is a method that changes something. It may be either the current object that is changed, or a parameter.

State

The stored data that a system accesses and alters; in particular, in object-oriented terms, the values of an object's properties (attributes), which can vary over its lifetime.

5.14. Bibliography

[ABR 96] ABRIAL J-R., *The B-Book: Assigning Programs to Meanings*, Cambridge University Press, 1996.

[ANS 89] American National Standard for Information Systems - Programming Language C, ANSI X3.159-1989, American National Standards Institute, 1989.

[BAR 03] BARNES J., *High Integrity Software: The SPARK Approach to Safety and Security*, Addison Wesley, 2003.

[BAR 04] BARNETT M., LEINO K.R.M., SCHULTE W., "The spec# programming system: an overview", *CASSIS 2004, LNCS*, vol. 3362, Springer, 2004.

[BUN 06] BUNDY A., "A very mathematical dilemma", *The Computer Journal*, vol. 49, no. 4, 2006.

[CRO 04a] CROCKER D., "Safe object-oriented software: the verified design-by-contract paradigm", *Proceedings of the Twelfth Safety-Critical Systems Symposium*, REDMILL F., ANDERSON T. (eds), p. 19-41, Springer-Verlag, London, 2004.

[CRO 04b] CROCKER D., CARLTON J., "A high productivity tool for formally-verified software development", *Originally Presented at Formal Methods Workshop in Industrial Applications*, Ghent (2003) and later updated (Escher Technologies, Available at http://www.eschertech.com/papers/pdpaper.pdf), 2004.

[CRO 07] CROCKER D., CARLTON J., "Verification of C programs using automated reasoning", *Presented at SEFM 2007, 5th IEEE International Conference on Software Engineering and Formal Methods*, London, UK, September 2007.

[CRO 09a] CROCKER D.,"Some Common C/C++ Coding Errors and Their Occurrence Rates", *Safety Systems*, vol. 18, no. 2, January 2009.

[CRO 09b] CROCKER D., MASON G., *Full press release available at: http://www.eschertech.com/news/gduk/PR_2009oct15.php*, October 2009.

[CRO 12] CROWCON DETECTION INSTRUMENTS LTD, *Crowcon Overview: http://www.crowcon.com/pages/profile.php*, accessed January 2012.

[COU 77] COUSOT P., COUSOT R., *Abstract Interpretation: A Unified Lattice Model for Static Analysis of Programs by Construction or Approximation of Fixpoints*, ACM POPL, 1977.

[DIJ 76] DIJKSTRA E.W., *A Discipline of Programming*, Prentice Hall, 1976.

[ECV 11] Escher Technologies Ltd, *Escher C Verifier Language Reference Manual* current version available at http://www.eschertech.com/product_documentation, 2011.

[EVA 02] EVANS D., LAROCHELLE D., "Improving security using extensible lightweight static analysis", *IEEE Software*, January/February 2002.

[FEI 08] FEINERER I., SALZER G., "A comparison of tools for teaching formal software verification", *Formal Aspects of Computing*, June 2008.

[FIL 03] FILLIÂTRE J.-C., *Why: A Multi-language Multi-prover Verification Tool*, Research Report 1366, LRI, University of Paris Sud, 2003.

[FIL 04] FILLIÂTRE J.-C., MARCHÉ C., "Multi-prover verification of C programs" *Lecture Notes in Computer Science 3308*, p. 15-29, 2004.

[FLO 67] FLOYD R.W., "Assigning meanings to programs", *Proceedings of the American Mathematical Society Symposia on Applied Mathematics*, vol. 19, p. 19-31, 1967.

[HOA 69] HOARE C.A.R., "An axiomatic basis for computer programming", *Communications of the ACM*, vol. 12, no. 10, p. 576-580, 583, October, 1969.

[ISO 99] *ISO/IEC 9899-1999, Programming Languages - C*, International Organization for Standardization, 1999.

[LEI 00] LEINO K.R.M., NELSON G., SAXE J.B., ESC/Java User's Manual Technical Note 2000-002, *Compaq Systems Research Center*, October 2000.

[LEI 05] LEINO K.R.M., FRANCESCO L., *Loop Invariants on Demand* APLAS 2005, p. 119-134, 2005.

[LIS 94] LISKOV B., WING J., "A behavioral notion of subtyping", *ACM Transactions on Programming Languages and Systems (TOPLAS)*, vol. 16, no. 6, p. 1811-1841, November 1994.

[MEY 88] MEYER B., *Object-Oriented Software Construction*, first edition, 1988.

[MIS 04] MISRA-C:2004 - *Guidelines for the Use of the C Language in Critical Systems, Motor Industry Software Reliability Association*, 2004.

[MUL 12] MULLINS T., Email, 25 January 2012, Griffith College, Dublin.

[PDL 11] Escher Technologies Ltd, *Perfect Developer Language Reference Manual: current version available at http://www.eschertech.com/product_documentation*, 2011.

[STA 99] STARK J., IRELAND A., "Invariant discovery via failed proof attempts", *Lecture Notes in Computer Science 1559*, p. 271-288, 1999.

[WAR 12] WARREN J., *Emails, 1 February 2012*, Precision Design Technology Ltd.

[ZHA 06] ZHAI C., LAFFERTY J., "A risk minimization framework for information retrieval", *Information Processing Management*, vol. 42, no. 1, p. 31-55, Pergamon Press, Inc., 2006.

Chapter 6

Partial Applications of Formal Methods

6.1. History

The AeS Group is a Brazilian company, established in 1991. At that time, it dealt with building automation. By 1998, it began its involvement in the railway sector when the first Brazilian General Door Control System (GDC) for rolling stock doors was developed. The aim of this equipment was to be an interface between train requests (operator request, signaling requests, etc.) and the door system itself. In this short chapter, its role is critical as the major operation of this equipment consists of sending an open command to the doors in each cabin. From that time, several safety-critical or safety-related systems were developed by AeS, such as "dead man control", parts of brake system, speed limit control, etc.

Due to the advances in technology, many safety functions handled earlier by hardware are now the responsibility of embedded software. Some standards can be followed to increase the equipment safety level. One of the most widely used is the IEC 61508 [EHA 07]. This standard presents four levels of safety (higher level, higher safety), the so-called Safety Integrity Levels – SIL. Above SIL2, the use of the formal method is required or suggested to achieve a certain level of completeness, robustness, and safety, which increases as the level becomes higher. This is a general standard that can be applied in several fields, and it is deeply related to the concept of *functional safety*, which means that the safety property is related to a function (or behavior) of the systems and not to a simple component (as it is done when reliability is the point of study). This is also a base standard for many others, among them, the

---

Chapter written by Aryldo G. RUSSO Jr.

CENELEC EN 50128, that is specific for the railway domain, and follows the same principle, although the safety classification is slightly different.

The goal of using formal methods is to produce an unambiguous and consistent specification that is as complete, error-free, and with as few contradictions as possible. However, it must be simple to verify.

Since 2006, AeS began its studies about formal methods to decide which method to use, where to apply it, how to apply it, etc. Later on, AeS became a part of the DEPLOY project as Deploy Associate. This chapter outlines case studies related to the project.

6.1.1. *Context*

Before going through the studies, a brief description of the AeS structure, projects, and team is needed to create a context for our applications:

– our Research and Development team is composed of 10 engineers, none of them with a formal method background, and one R&D manager (engineer with some formal background);

– during the time of the DEPLOY project, the team received a whole week's training covering the Rodin environment, refinement, and UML-B;

– as AeS is inserted in the railway domain, several standards need to be followed, some of them related to safety, such as EN 50128 and IEC 61508. Those standards recommend the use of formal methods for application, which require higher levels of safety (SIL3 and SIL4).

6.2. Case studies

In this section, we present a chronological description of our case studies, trying to show what the problem was, what we applied, and what the results were. One important point is that in none of our case studies did we use formal methods (or formal methodologies) from the beginning to the end of the development process. We only applied it precisely at the point that we thought it would be effective, avoiding extra costs and efforts. Nevertheless, we think, some time in the future, a complete formalized development process might be used and evaluated in comparison with the traditional approach. This is the last case study we are carrying out, but unfortunately, it was not finished at the time of writing.

6.2.1. *Early 2007 – The B method and railway domain – breaking the wall*

During our search of formal methods (that began in 2006), we found that the B method had become popular in the railway domain. At that time, the reason was not evident, and we decided to follow this flow. Only theoretical publications could be found about the subject and no "easy-to-use guide" was available. We learned that a conference about B was to take place in 2007 in Besanson, France. This conference was our first contact with the real world of academic development. It was also our first contact with people committed to deploying the B/Event-B method and with the Rodin tool. Moreover, it was the first contact with our first case study – the bridge between continent and isle. And this was the reason for us to now be a Deploy associate partner, and to successfully use formal methods in our applications.

Although this was our first contact, several attempts on the use of formal methods had already been made before, and, some of them resulted in successful outcomes, as can be seen at [LEC 07]. Of course, we could, from the beginning, try the application as well, and here is our first attempt [EHA 07].

6.2.2. *Early 2008 – requirements verification*

In our earlier attempts to use formal methods, we tried to apply it at the beginning of the development process, namely, in the process of converting the natural language specification into a formal one. We could not find a tool that would help us to do it manually, but after studying the formalization process and the notion of abstraction and refinement, this helped in changing the way of thinking, improving our understanding of the specification.

As a result of this initial attempt, I can cite two small (in terms of size) but important achievements:

– we could find natural language specification inconsistencies: we formalized a small portion of a door system specification, and, at that point our objective was to formalize exactly what was written in that specification. Doing so we reached a point where the formal abstraction could not be proved, and analyzing the cause, we found contradictory information imported from the natural language specification;

– we could find missing information on natural language specification. We tried to formalize the existing specification of our pilot project: a small system that should stop the movement of the train in case of human problems with the operator. This was a one-sheet only specification. As in the former case, we formalize exactly what we had in the natural language specification and we could discover that there were a lot of missing information. This missing points lead us to several assumptions that shall be validated with the customer specialist. Only after this validation and changing of the natural language specification the formal model could be finished.

At the end, in both attempts, we realized that we could use formalization as a tool to verify and correct customers' specifications.

6.2.2.1. *Example*

In this small but actual example, we performed an exercise, formalizing a specification, and we realized that there was a lot information that was missing. We had to guess (or to request clarification from the customer) what the possible options were to fulfill the gaps in the specification. This exercise helped in changing the way we built the specification, using the formal language as a tool to find these gaps.

This system is called "dead man control" and the basic requirements are as follows: (the full requirements elicitation is also part of this research):

1) if the train is in automatic mode, the dead man control must be disabled;

2) if the train is in manual mode, the dead man control must be enabled;

3) in manual mode the operator must push the control button every X seconds;

4) after X seconds; if the operator has not pressed the button, the system must provide an alarm sound;

5) after X seconds after the alarm activation, if the operator has not pressed the button, the system must stop the train;

6) whenever the operator presses the button, the system must go back to the normal situation.

Of course, it is just an example. A formal specification of the door system has more than 7,000 of event-B lines (only for the abstract model).

Another example was used to find a contradiction. The full specification in natural language is more that 100 pages long, and the information is spread out. There were some contradictions though, which were difficult to find by just reading the specification. But, when you try to put it formally, you can see that there are some problems.

Here, you can find the specification based on the natural language document. Where's the mistake?

MACHINE Open_contradiction

VARIABLES

`train_stopped` boolean. when the train is stopped its value is TRUE

`train_low_speed` boolean. when the train speed is below 6 km/h its value is TRUE

`door_authorization` boolean. when the train is allowed to open doors its value is TRUE

`emergency_button` boolean. if the button is pressed, its value is TRUE `open_command` boolean. if true, command the opening

`train_speed` NAT. real speed

INVARIANTS

inv1 : $train\_stopped \in BOOL$

inv2 : $door\_authorization \in BOOL$

inv3 : $train\_low\_speed \in BOOL$

inv4 : $emergency\_button \in BOOL$

inv5 : $train\_stopped = TRUE \Rightarrow door\_authorization = TRUE$

inv6 : $train\_stopped = FALSE \Rightarrow door\_authorization = FALSE$

inv7 : $train\_stopped = TRUE \Rightarrow train\_low\_speed = TRUE$

inv9 : $open\_command \in BOOL$

inv10 : $train\_speed \in \mathbb{N}$

inv11 : $door\_authorization = FALSE \vee (\neg(train\_low\_speed = TRUE and$
$emergency\_button = TRUE)) \Rightarrow open\_command = FALSE$

EVENTS

Initialization

begin

act1 : $door\_authorization := TRUE$

act2 : $train\_stopped := TRUE$

act3 : $train\_low\_speed := TRUE$

act4 : $emergency\_button := FALSE$

act5 : $open\_command := FALSE$

act6 : $train\_speed := 0$

end

Event *EMERGENCY_OPEN* $\widehat{=}$

when

grd1 : $train\_low\_speed = TRUE$

grd3 : $emergency\_button = TRUE$

then

act1 : $open\_command := TRUE$

end

Event *LOW_SPEED_MONITOR* $\widehat{=}$

when

grd1 : $train\_speed \leq 6$

then

act1 : $train\_low\_speed := TRUE$

end

Event *ZERO_SPEED_MONITOR* $\widehat{=}$

when

grd1 : $train\_speed = 0$

then

act1 : $train\_stopped := TRUE$

act2 : $train\_low\_speed := TRUE$

act3 : $door\_authorization := TRUE$

end

Event *AUTHORIZATION_RELEASE* $\widehat{=}$

when

grd1 : $train\_speed > 0$

then

act1 : $door\_authorization := FALSE$

act2 : $train\_stopped := FALSE$

act3 : $open\_command := FALSE$

end

Event *LOW_SPEED_RELEASE* $\widehat{=}$

when

grd1 : $train\_speed > 6$

then

act1 : $train\_low\_speed := FALSE$

act2 : $train\_stopped := FALSE$

act3 : $door\_authorization := FALSE$

act4 : $open\_command := FALSE$

end

END

6.2.3. *2009 – tool comparison*

To verify how the current tools can be modified to reflect industrial needs, I prepared a brief comparison of some existing tools. I have restricted this comparison to some tools that I knew well and that have been used in my application field, that is in railways application. Those tools are Atelier B, Rodin, and SCADE[1]. This study was performed three years ago, so some subsequent changes are possible at this time.

6.2.3.1. *Methodology*

As this was not a scientific comparison, the way that I choose to determine the classification of each tool in each category was my personal feeling while adopting those tools in our development process.

It is also prudent to state the maturity differences among these tools. While SCADE and Atelier B have been in the market for a long time, Rodin was about to be released in its first official version (version 1.0). This means that the first two tools have already received a lot of feedback from their industrial users, helping them to change the direction when users were not satisfied (it was the Atelier B case, where after a lot of complaints about the user interface, its GUI was changed completely), while the last one has not yet received or implemented completely this feedback.

The comparison methodology was based on three aspects, as follows:

– *capability*: in this case, we try to verify how these tools can satisfy project constraints;

– *usability*: basically, what's the difficulty the user faces when trying to use the tool;

– *adequacy to the current development process*: that is, how the tool can better fit in the process without causing too many changes in the way it has performed so far.

To make a classification of these aspects I used a simple ranking method, as follows:

– 1 Very difficult;

– 2 Medium;

– 3 Easy.

We present the results in Table 6.1.

1 Versions: Atelier B 3.7, SCADE 6.0, and Rodin 0.9.

Aspect	**Capability**	**Usability**	**Adaptation**	***Results***
AtelierB	2	1	2	*5*
Rodin	2	2	1	*5*
SCADE	2	3	3	*8*

Table 6.1. *Comparison table*

6.2.3.2. *Chart comparison*

To justify those results, I can say that:

– Atelier B:

- the capability to solve the project constraints is not so bad, but you do need to know a lot of the formal language and constructs to be able to have easy proof obligations,

- although version 4 of Atelier B demonstrates better usability, all comments I have so far are based on the previous version where the lack of a good User Interface makes its difficult to use,

- since it allows us to go from the specification to the code, it can be considered as a good tool for that purpose, but as the interactions during the middle phases (refinements) are some times cumbersome, it cannot receive the higher grade;

– Rodin:

- as it is not very different from Atelier B, similar results could be seen. The capability to solve the project constraints is quite good, but you do need to know a lot of the formal language and constructs to be able to have easy proof obligations,

- the way that Rodin was constructed helps an inexperienced person since you just need to fill in some fields to have a basic specification. However, the absence of a text editor that could help a more experienced person and speed up the specification process lowers its classification,

- as, at the moment of the evaluation, there were no possibilities of decomposition, and the ability to help only in the system specification phase, turn it in a difficult tool to be used in the current process;

– SCADE:

- even based on a different concept, where formal methods are behind the scene, it has a great capability to deal with project constraints, but you still need some formal background to construct correct models,

- as it was built from the very beginning to be an industrial tool, its usability is its strongest point, with a good interface and a lot of fancy features that captivate the user. A lot of things can be done based on templates and patterns, what helps a lot as well,

- besides the capability to go from the specification to the code, it has also some other complementary tools that help you in important auxiliary tasks in the project, such as requirement management, traceability, etc.

6.2.4. *Late 2009 – writing a formal specification – user point of view*

B [ABR 96] is a formal method that allows us to produce proof obligations that demonstrate correctness of the design and the refinement. Nevertheless, there is no standard mechanism for mapping requirements to formal specifications. To overcome this issue, different solutions have been proposed by researchers. In [PON 06], the authors have presented traceability between KAOS requirements and B. Some authors are investigating the use of the Problem Frames approach [JAC 01] as a possible response. A mixed solution using natural language and UML-B has been proposed by [JAS 09]. However, these approaches use non-standard artifacts for requirement specifications, which are a disincentive for designers to adopt formal methods since they must spend time to learn them.

Use cases [JAC 92] can be considered as the *de facto* industry standard for requirement specifications. They provide a good way to capture how the end user interacts with the system by detailing scenario-driven threads. A typical use case describes a user-valued transaction in a sequence of steps expressed in a natural language, which makes use cases readable for most end-users. In [LED] the author has presented an approach for building B specifications from use-case models. This approach is similar to ours, but his project has focused on bringing the object-oriented paradigm (including UML diagrams) to formal methods. His method also maps each use case as a unique B operation, which we believe is incorrect since each use case can have many transactions according to Jacobson [JAC 87].

Another relevant point about use cases is the possibility to derive test cases. One of the new prominent development models is the so-called Test Driven Development [BEC 02], where the input information used to generate the source code is the test case, instead of use case scenarios or other traditional requirement documentation.

We tried an approach for starting B specifications from use and test cases. Use case transaction identification can be used as a guideline for defining B operations, including the pre- and post-conditions. In the same way, test cases can help in the definition of global invariants and constraints.

6.2.4.1. *Mapping UC to B*

The aim of this approach is to fulfill some intermediate phases in the proposed development process. So far, the use of formal methods in the development process is only from the design phase where the requirements are already well defined, but this is not the case with industrial projects.

Based on that, the method proposed here would be useful during the early phases, to help in both directions:

– to the top, helping the requirements elicitation (what might be combined with other techniques and methods, such as Jackson's Problem Frames);

– to the bottom, helping the creation of the first abstract formal model, in the sense that it can support the first definitions.

For mapping use cases to traditional B specifications, we propose that use case scenario sentences must be written using a controlled natural language (CNL) described according to our use case transaction definition, which is based on Ochodek's transaction model [OCH 08].

DEFINITION 6.1.– A transaction is the shortest sequence of an actor's and a system actions, which starts from the actors request and finishes with the system response. The system validation and system expletive actions must also occur within the starting and ending action. The pattern for a transaction written as a sequence of four steps in a scenario:

n. An actor's request action (U).

n+1. A system data validation action (SV).

n+2. A system expletive action (e.g. system state change action) (SE).

n+3. A system response action (SR).

We have also decided to define the grammar using subject-verb-object (SVO) sentences because they are good at telling the sequence of events. We have mapped the use-case actor as subject, a set of action's predicate synonyms (for example, validate, verify etc. would be grouped together) as verb, and the rest of the sentence as object.

For the sake of simplicity, we assume that all environmental tasks, like user interactions, are perfect, i.e. they are always executed in a correct order and at the right time. Further studies will be developed to incorporate non-functional requirements and exceptions.

Let's take a look at an example to clarify our idea. In accordance with the definition (6.1), suppose we have the following use case scenario:

```
1. The agent specifies a travel itinerary for a client.
2. The system validates the itinerary.
3. The system searches for a set of appropriate
   flights.
4. The system presents them to the agent.
5. The agent selects a flight.
6. The system verifies free spaces on the flight.
7. The system reserves any seat from the set of free
   spaces.
8. The system confirms the reservation.
```

In the above example, we can see two transactions (1-4 and 5-8) and each one can be used as a guide to create a B operation. From SV actions (2 and 6) we extract the preconditions (validates the itinerary, verifies free spaces) and from SE actions, (3 and 7) we derive the operations names (search, reserve) and the post-conditions (searches a set of flights, reserves any seat). A possible mapping to B specifications, which is self-explanatory, is shown below.

flights $\leftarrow$ search(city1,city2) $\triangleq$
  PRE city1 $\in$ CITY $\wedge$ city1 $\in$ CITY $\wedge$ city2 $\in$ itinerary(city1) **(2)**
  THEN flights := companies({city1 $\mapsto$ city2}) **(3)**
  END;

reserve(flight) $\triangleq$
  PRE card(flight) > 0 **(6)**
  THEN ANY seat **WHERE** seat $\in$ flight **(7)**
    THEN flight := flight - {seat}
    END
  END

6.2.4.2. *Results*

We used an approach for mapping requirements to B (Event B) models from use cases and test cases. We were not interested (at this moment) in the automatic translation of use cases for formal specifications since there were many natural language ambiguity problems. The intention was to take the use cases as a guideline for starting B specifications.

This was only an attempt to find out how well formal methods could fit in a UML development process. A lot of work has to be undertaken to make it a mature process.

6.2.5. *Early 2010 – a methodological approach to a B formalization*

Developing a formal specification of a system from the informal functional requirements of a customer remains an open issue in the software engineering world. Good requirements, which are well written and easy to understand, are essential [JAC 01] but not sufficient for achieving this goal.

We propose a framework to structure and organize such requirements, resulting in a requirements specification document. This Meta-model facilitates the development of a formal abstract model of the system functionality using the B method.

In this work, requirements elicitation is done without feedback from the customer, for two reasons. First, requirements elicitation often produces inconsistencies and redundancies. One needs to produce a formal representation that presents the requirements to the customer, beyond any doubt; so, we will consider the customer feedback once we have constructed the model and noted the inconsistencies. The lack of information has to be clearly pointed out and proved. Second, this work is part of a larger project to develop a method for methodically organizing the requirements of a system having only natural language documents as input. As an offshoot of this work, the formal model could help the customer to rewrite the natural language document to resolve such inconsistencies and redundancies to future projects with similar requirements.

To reduce the gap between the requirements and the formal model of the system and to come through with a clearer traceability, it is helpful to use an intermediate representation such as use cases [JR 09], problem frames [JAC 01], or KAOS [DAR 97]. This work employs an adaptation of the WRSPM approach [GUN 00], a reference model that can be adapted without requiring a whole new language to represent the requirements (as opposed to KAOS or problem frames). This is the basis of the method explored in this work.

To achieve a model that we can verify and validate, which points out doubtless inconsistencies, redundancies, and lack of information, the B method is applied over the specification from WRSPM. Once a formal model is constructed, one can use animation tools such as ProB to show this to the customer. They would be much easier to understand than mathematical obligation proofs. It is also in this process that we can verify the advantages of the methodology and identify the main complications of developing an abstract formal model using the horizontal refinements approach, which can lead to alternatives to deal with this issue in future work.

6.2.5.1. *The methodology*

The method we propose to use to generate the WRSPM-based model consists of taking the customer requirements (expressed in natural language) one-by-one. This way, we incrementally produce the specification **S** in terms of *phenomena*, *invariants*, and *operations*. We break the requirements into **W**, **R**, and establish traceability relating the labels of the *atomic* requirements with the labels of the invariants and operations manually. In future, with the development of tools, this should be automatic.

In Figure 6.1 we have a fluxogram illustrating the process step-by-step. We take it to each requirement, (already decomposed into an *atomic* requirement) and firstly identify all the *phenomena*. Then, to each *phenomenon* that is being identified for the first time we define the invariants with reference to its type and next the invariants and operations with reference to its state. Once all *phenomena* to that single requirement are identified and (the new ones) typed and invariant/operation related, we break the text of the requirement into information of the environment (**W**) and constraints of the system (**R**). The last step is to provide the traceability, making the links between the labels of the *atomic* requirements and the elements of the WRSPM specification.

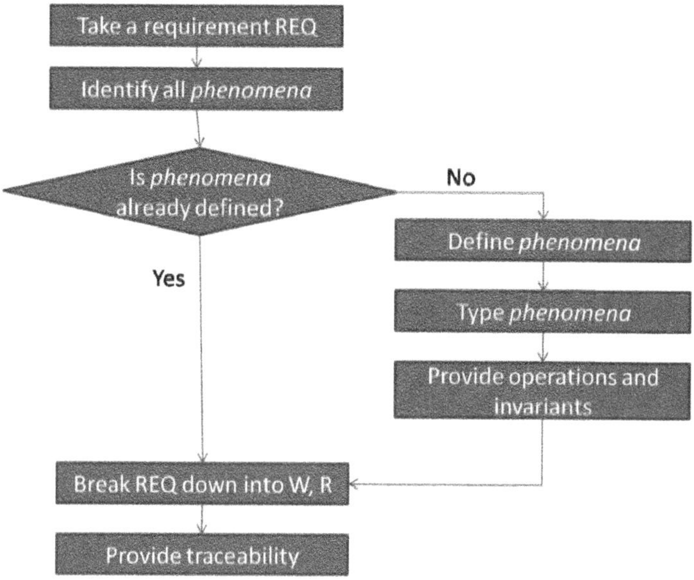

Figure 6.1. *Fluxogram illustrating the processing of requirements*

To model a complete system this way, we need to decide on a formal language that is rich enough to express the formal artifacts and phenomena.

Once all requirements are thus modeled, we have a complete abstract formal representation of the system (using the B method, for example). This is the starting point of a series of refinements, the last one resulting in **P** (in some programming language), to be executed in some execution environment **M**. In the process, dependency links shall be maintained between all artifacts to guarantee traceability. Upon completion of these steps, we have a correct and complete implementation of the system (since the requirements are also correct and complete).

6.2.5.2. *Horizontal refinements*

Some systems are too large and complex to have their formal modeling (the abstract model) done in just one step, and are more predicated to have it done in successive steps. That's where the idea of *horizontal refinements* (or superposition) fits in.

The idea consists of starting with an abstract model that only considers part of the requirements, and then introduces new requirements through successive refinements, adding new variables, strengthening preconditions, adding invariants, actions, or operations.

The one doing the modeling process has to explore the requirements document (in the methodology presented in this chapter, the WRSPM model) and gradually take from it the elements to be formalized.

It is important to take care to, correctly, define the architecture that supports the incremental addition of information, and to define a good order to add these requirements, so abstraction disparities are not faced (like a very concrete requirement being added in a highly abstract level).

Another issue is to guarantee completeness with the previous model (WRSPM): the horizontal refinements process must last until all the requirements and definitions have been taken into account, and only after this phase is over, we will start the process of the vertical refinements, pushing the model into more concrete levels through design decisions until a level concrete enough to generate code is reached.

6.2.5.3. *Lessons learned*

Once we had added what we considered enough to the abstract model (to be further expanded with the horizontal refinements), this model was analyzed using the animations tool ProB [LEU 03], which provides features such as model checking and execution step-by-step to achieve some state through the operations of the specification. This allowed us to explore the B specification, checking for errors and identifying inconsistencies and lack of information. Undefined states to the variables were reached since we did not know, guided by the customer requirements, what to do in certain conditions. It would also be possible to show the customer the result of the project so far, providing the ProB examples of the model execution.

The next step would be to refine horizontally the abstract model of the system. Earlier, we only considered part of the requirements in the process of the construction of the model, so the new requirements had to be introduced through refinements. That is where we had big conceptual problems.

To model the architecture, communication, and complexity decomposition with the B Method, the clause **INCLUDES** is necessary, and this clause cannot be used to include a *refinement*, which means, in a practical example: the *machine System* of the abstract model cannot include the *refinement CPG_r*, which cannot include the *refinement Opening_r*, and so on. Besides this, one cannot add a new operation through refinements. It is necessary to go back to the abstract model, define the whole signature of the operation (with the variables to be received and returned) and perform the refinement step, which clearly destroys the concept of *expanding* the information. In order to have an abstract model able to be horizontally refined one would have to do an analysis through *all* the requirements to define at least the assignature of *all* the operations that are going to be performed in *all* the machines. Therefore, to apply horizontal refinements, we decided to use another approach to build the model and perform the refinements, using the Event-B formal methodology and its extensions that enables the decomposition process.

In Event-B we are allowed to add events (analog to operations in B) through refinements, so we can *expand* the model without cheating (like with the B method). The key point to the horizontal refinements in Event-B is, however, the *decomposition*; By default, **INCLUDES** abstraction, there is no communication between different *machines*, but one can introduce it through *decompositions*, based on the concepts of *shared events* (the same event is shared by independent *machines*, which models the architecture communication between different components) and *shared variables* (the same variable is shared by independent *machines*, which models complexity division inside the same component).

Using this feature, we will be able to model the architecture communication (through shared events) between the CGP and the environment, and model the complexity division (through shared variables) between the CGP and its nested components. Once the decomposition is done, we can independently refine the decomposed *machines*, getting a more expanded model and doing more decompositions if necessary.

6.2.6. *Early 2011 changing the way to vital verification*

We present the architecture and approach of a graphical tool prototype based on formal methods to validate tracks topologies and train movement conditions. This tool is Eclipse-based and comes with the Rodin platform. Its main features are summed up in the graphical simulation of railway specifications and the train movement properties' validation. The tool is called VeRaSiS and uses Event-B to formalize, prove, and verify the generated properties.

In the context of safety-critical applications, where failures in the system can potentially lead to life-threatening issues or huge financial losses, such as in nuclear, medical, or railway domains, some properties need to be exhaustively verified to guarantee a minimum level of confidence in the system. Exhaustive verification aims at ensuring that the system will not allow dangerous situations or, if such situations happen, that they will activate the right protection to avoid its propagation. When talking about the railway domain, one of the most important safety-critical system applications is considered to be the signaling system, which is in charge of, among other tasks, dictating the train movement in a specific direction and secured condition. To guarantee these conditions, the track topology and the movement conditions are usually manually translated into a set of properties that are, after completion, verified and validated through the expert eyes of the validator. This process chain does not seem sufficiently reliable to guarantee that errors and/or inconsistencies that are very difficult to pinpoint during the validation phase will not creep into the writing process. Nowadays, the superior effectiveness of formal methods over manual validation in ensuring safety in critical systems has been clearly demonstrated. Yet, despite the volumes of research and tools developed to facilitate the use of formal methods, mainstream developers still have a certain reluctance to use these methods. To help those developers overcome such a resistance, this chapter introduces an automatic tool to generate and validate the safety properties of the trains movement. This tool is called VeRaSiS, and uses Event-B to formalize, prove, and verify the generated properties.

In the mid-1980s, Siemens Transportation Systems (STS), now called SIMENS mobility, (formerly called Matra Transport International), was the first company to lead a subway line automation project. This concept has been of great difficulty because of implications raised on human life. Up until then, the usual development process was not secure enough to cope with the issue of the safety of subway passengers. The need to increase the confidence in the development process has led to the use of formal methods. Line 14 of the Paris Metro is one of the most exemplary outcomes of this project. Currently, the line carries more than 45,000 passengers per day and since then several automatic rail lines have been built [GUN 00]. Nowadays, STS and some other companies have become experts at using formal methods in railway requirements. However, mainstream software developers have not been completely successful in mastering the use of these methods. The use of formal methods in the railway domain depends mainly on the requirements type. This focuses on the trains movement. More precisely, it takes a close look at the Boolean equation validation issue.

6.2.6.1. *The Boolean equations validation problem*

Boolean equations or Boolean logic is a logical calculus of truth values. It has many applications in electronics, computer hardware, and is the basis of all modern digital electronics. With reference to the railway sector Boolean equations are used to express safety moving properties. In other words, they define properties enabling the movement from one point to another in a railway system. A set of Boolean equations has to be generated for all paths in the model designed. It is called the Boolean Equation Library (BEL). In industrial projects, those equations are generated manually and validated either manually or with formal tools depending on companies. Besides the time spent to construct and validate these equations, sometimes errors or inconsistencies can still creep into them after the manual validation has been done, a probability which can, at worst, lead to accidents or, at best, if detected on time (that is, during the test phase), force developers to rework the equations from scratch.

6.2.6.2. *Using formal methods to validate Boolean equations*

A preliminary analysis using a confidential real case provided by a Brazilian transportation company was made in order to, firstly, test the efficiently use of formal methods, and secondly, to convince local industrials to change the manual validation to a formal validation. This study aimed to revalidate a specification, which was already validated manually. First, the example has been translated into a classical B machine. A parser converting the Boolean equations to B [JAC 01] notation has been developed to automatize the translation. This machine has been run in Prob and as a result, it has given a counterexample showing small errors, which were not detected in the first validation. Two speed limits could be selected, non-deterministically, in a moment. This case study has been taken at random, which means that probably there are other examples that contain errors and which were validated manually. This study has proven to industrialists the necessity to supersede the manual validation by a formal validation. It has also revealed the necessity of developing a formal tool to automatize the Boolean equations validation based on in a model graphically designed.

6.2.7. *The VeRaSiS plug-in*

The VeRaSiS plug-in is a tool allowing validation of the trains movement in a railway system. It is Eclipse-based and is built for extensibility. The VeRaSiS plug-in proposes three main features: firstly, a graphical simulation of use cases; secondly, a fully automatic Boolean equations generation, and finally the Boolean equations validation. The VeRaSiS plug-in, see Figure 6.2, comes up with a graphical interface allowing users to design railways systems. The BEL is automatically generated. After this, a formal tool is used to validate it. Error messages are translated into straightforward messages to keep users outside the complicated formal language involved. That said, a minimum contact with formalism is needed to be used in industrial environment.

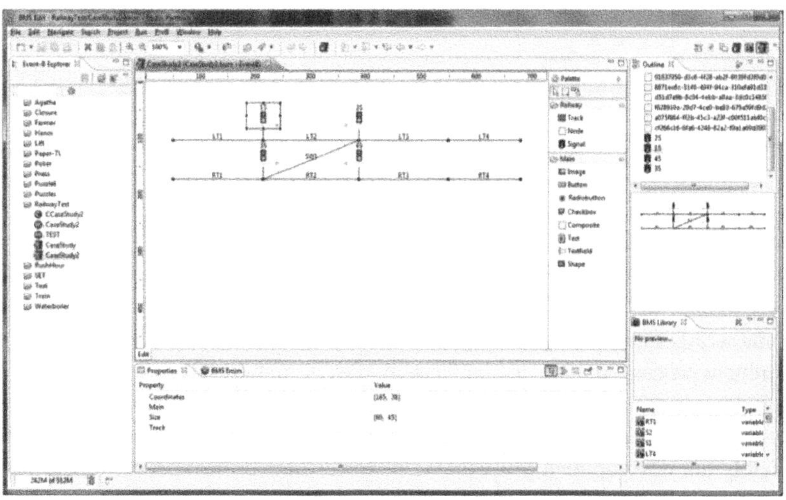

Figure 6.2. *VeRaSiS plug-in*

6.2.8. *Results*

We have presented a new formal tool to model railway specifications and validate train movement properties. This Eclipse plug-in is designed for industrial needs and developed to deal with the challenge that mainstream users can develop with formal railway models while remaining as far as possible from the difficulties of formal specifications. The VeRaSiS plug-in aims to replace the traditional Boolean equations validation process, which is not strong enough to guarantee the safety level required as stated before – a risk that upset industrialists and the reason why they have given the green signal to start this project. In addition to the tools that we have to develop to perform the VeRaSiS plug-in, this study proves the VeRaSiS concept in a case study based on a industrial specification. That said, it reveals that we have to optimize the algorithm allowing us to generate the BEL from the model designed and the translation template from BEL to Event-B. In the case study that we are working on right now, normally the translation happens straightforward in a one-to-one relation, as one BEL becomes one event. However as the complexity grows, introducing new a kind of BEL, this rule might not be valid anymore, which implies that there are further studies that need to be done. We also need to generate a set of rules for each kind of BEL that needs to be created in a real project, as in the case study we selected only one (or two) kinds of BELs, namely the path allowing us to validate the concept, but all equations needed have to be generated in a operational model. Additional case studies would certainly find other limitations of the VeRaSiS concepts. For instance our main aim is to have a fully operating first version of the VeRaSiS plugin in order to prove its functionality.

6.3. Conclusion

In this chapter, we summarize what we have been doing so far in the context of the DEPLOY Associate program. Besides the progress of the pilot project, the development of a specification of a simple railway system called dead man control, we also present some parallel developments, some theoretical, some practical, of the use of formal methods in industry. Instead of presenting the theoretical aspects of each project, we present here a description of the lessons learned in each project focusing on some important points, such as requirement validation, productivity, and dependability.

Some words should be mentioned about the introduction of formal methods in the development process, mainly, in small companies (as in big ones, it seems to be easy as you could just hire some specialists and get them to work together with the development team). At the moment, even after some training, we did not find a good way to change completely our development process. We do have some trained people that are able to understand and use, with some restrictions, event-B and Rodin. However, it takes a lot of time to get rid of the traditional approach and establish a new one (moreover when all products are already under development).

To overcome this problem, we introduced FM in parts, in points where (we thought) it would be of some help to improve the product quality and dependability. Still it is not a common procedure, but we could see some advantages. In a perfect world, the approach of parallel development (FM and traditional) would be perfect to convince people of its power and to create a new methodology for the development process to substitute the old one, but this is not a perfect world.

Although using formal methods informally does really help in changing the way of thinking, meaning that, even if not all the process is formalized, the people involved in the development are following a refinement process.

So far we use formal methods in different ways, such as:

1) Verify gaps and mistakes in the NL requirements. We modeled the specified system (only the abstract model) and used it to show some inconsistencies in the NL spec. There is a tremendous gap between the requirement and the formal model. This is a problem in big industrial projects as the main task of the management team is to show that what was developed met what was required. We did not find a way to do that at this point (although we are closely related to the ProR project trying to see if we can solve this problem). Again, it is not that difficult to maintain this traceability, informally or internally, but in real projects, it needs to be transparent to all involved.

2) We have specified, using different approaches, the door system (B, Event-B, UML-B, etc.). In this, I can talk about re-use. This system is a kind of distributed one, and in different projects, different parts of the "standard" system might be used, you could consider it as a product line, if you want. We did not find a way to re-use "parts" of the specification, only the complete one, altering what is necessary to meet the new requirements in a new project. To exemplify, this system is divided into three types of equipment, which may or may not be used during the project. Even with the decomposition tool, it is not clear whether it would be possible to re-use parts of the specification.

3) Regarding dependability, this will need future work. I do think it increases the dependability of the system, although I have no strong evidence to prove that. Without code generation, it is not clear if the mistakes that were introduced earlier in the specification step, will not be introduced in the coding process. The key point is that formalizing the system helps to gain a better understanding of it, and as a result prevent some errors from being introduced.

6.4. Bibliography

[ABR 96] ABRIAL J.-R., *The B-book: Assigning Programs to Meanings*, Cambridge University Press, New York, NY, USA, 1996.

[BEC 02] BECK K., *Test Driven Development: By Example*, Addison-Wesley Longman Publishing Company, Incorporated, Boston, MA, USA, 2002.

[DAR 97] DARIMONT R., DELOR E., MASSONET P., LAMSWEERDE A., "GRAIL/KAOS: an environment for goal-driven requirements engineering", *ICSE '97: Proceedings of the 19th International Conference on Software Engineering*, May 1997.

[EHA 07] EHARBE D.D., MOREIRA A.M., SILVA P. S.M., JR A.G.R., "Modelling control systems in B: an industrial case study", *SBMF 2007 - 10th Symposium on Formal Methods*, p. 16, July 2007.

[GUN 00] GUNTER C., GUNTER E., JACKSON M., ZAVE P., "A reference model for requirements and specifications", *IEEE SOFTWARE*, vol. 17, no. 3, p. 37-43, 2000.

[JAC 87] JACOBSON I., "Object-oriented development in an industrial environment", *OOPSLA '87: Conference Proceedings on Object-oriented Programming Systems, Languages and Applications*, New York, p. 183-191, NY, USA, ACM, 1987.

[JAC 92] JACOBSON I., *Object-oriented Software Engineering: a Use Case Driven Approach*, Addison-Wesley Professional, June 1992.

[JAC 01] JACKSON M., *Problem Frames: Analyzing and Structuring Software Development Problems*, Addison-Wesley Longman Publishing Company, Incorporated, Boston, MA, USA, 2001.

[JAS 09] JASTRAM M., LEUSCHEL M., BENDISPOSTO J., JR A.R., "Mapping requirements to B models", *Unpublished manuscript*, 2009.

[JR 09] JR A.R., DE SOUSA T., Starting B Specifications from Use Cases, 2009.

[LEC 07] LECOMTE T., SERVAT T., POUZANCRE G., "Formal methods in safety-critical railway systems", *Proceedings of Brazilian Symposium on Formal Methods: SMBF*, January 2007.

[LED] LEDANG H., *Automated Software Engineering Conference*.

[LEU 03] LEUSCHEL M., BUTLER M., "ProB: a model checker for B", p. 855-874, 2003.

[OCH 08] OCHODEK M., NAWROCKI J., *Automatic Transactions Identification in Use Cases*, Springer-Verlag, 2008.

[PON 06] PONSARD C., DIEUL E., "From requirements models to formal specifications in B", *ReMo2V CEUR Workshop Proceedings*, vol. 241, 2006.

Chapter 7

Event-B and Rodin

7.1. Event-B

7.1.1. *The Event-B definition*

Event-B [ABR 08] is a formal method for specifying, modeling, and reasoning about system based on set theory and predicate logic. Event-B evolved from Classical B [ABR 96] and Action Systems [BAC 89]. On the one hand, Event-B is a simplification as well as an evolution of the B-Methods; on the other hand, Event-B is influenced by the action system approach. It has the same structure as an action system, which describes the behavior of a reactive system in terms of the guarded actions that can take place during its execution.

Event-B is different from the B method in some aspects. The B method is organized in a way that is suitable for the development of non-concurrent programs, whereas Event-B is geared toward the development of systems including reactive and concurrent systems. Building a model in Event-B starts with a very abstract level, and continues in different abstraction levels by use of refinement, which will be explained in section 7.1.3. Event-B uses mathematical proof to verify consistency between refinement levels. Association of proof obligations in Event-B permits us to reason about it; see section 7.1.4. Rodin is a tool platform for modeling and proving in Event-B. It will be outlined in section 7.2. Section 7.4 describes the development of a metro system case study in Event-B.

Chapter written by Michael BUTLER, Asieh SALEHI FATHABADI and Renato SILVA.

7.1.2. *Event-B structure and notation*

A model in Event-B [ABR 08] consists of *contexts* and *machines*. Contexts contain the static part (types and constants) of a model while machines contain the dynamic part (variables and events). Contexts provide axiomatic properties of an Event-B model, whereas machines provide behavioral properties of an Event-B model. Items of machines and contexts presented in this section are called modeling elements. There are various relationships between contexts and machines. A context can be "extended" by other contexts and "seen" by machines. A machine can be "refined" by other machines and refer to contexts as its static part. Refinement is described more in section 7.1.3. The relationship between machine and context is illustrated in Figure 7.1.

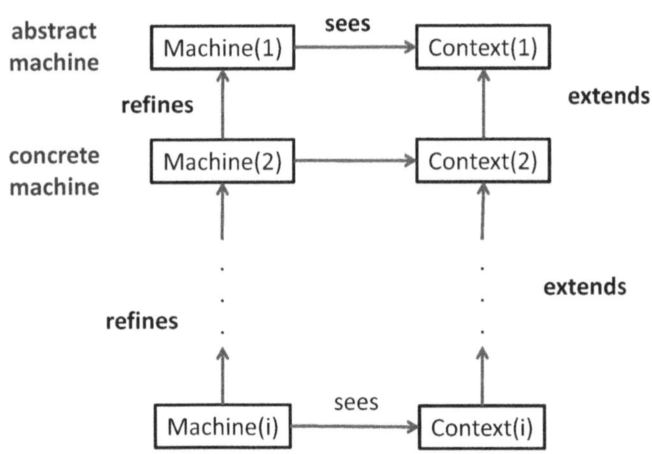

Figure 7.1. *Machine and context relationships*

From a given machine, $Machine1$ in this case, a new machine, $Machine2$, can be built as a refinement of $Machine1$. In this case, $Machine1$ is called an abstraction of $Machine2$, and $Machine2$ is said to be a concrete version of $Machine1$.

7.1.2.1. *Context structure*

Modeling elements of a context [ABR 08] are of four forms: sets, constants, axioms, theorems. This is illustrated in Figure 7.2. Axioms are predicates that describe the properties of sets and constants. Theorems are properties that should follow from the axioms. A context may extend one or more other contexts, and can also be seen by several machines in a direct or indirect way.

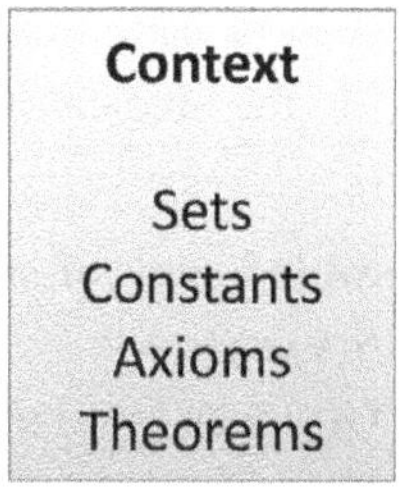

Figure 7.2. *Structure of a context*

7.1.2.2. *Machine Structure*

A machine [ABR 08] consists of variables, invariants, events, theorems, and variants, illustrated in Figure 7.3. Variables, v, define the states of a model. Invariants, $I(v)$, constrain variables, and are supposed to be "hold" whenever variables are changed by an event. New events can be defined in a concrete machine. These will be described more in section 7.1.3. To prove that they do not take control forever, a new event must decrease a natural number expression called a variant.

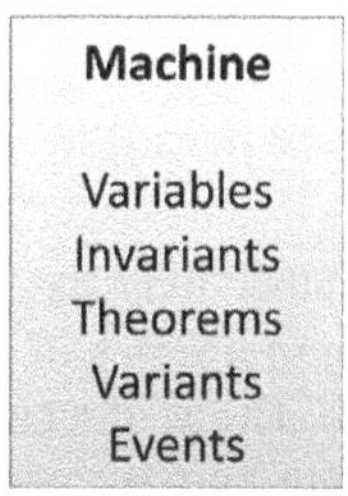

Figure 7.3. *Structure of a machine*

7.1.2.3. *Events*

In Event-B, state of a model is changed by means of event execution. Each event is composed of a name, a set of guards $G(t, v)$, and some actions $S(t, v)$, where t are parameters of the event and v is state of the system, which is defined by variables. All events are automatic and can be executed only when their guards hold. When the guards of several events hold at the same time, then only one of those events is chosen non-deterministically to be executed. An event can appear in three forms presented in Table 7.1. In the simplest terms, an event contains only some actions, in the second form it can be composed of guards and actions without parameters, and finally in the third form, an event has guards, actions, and some parameters.

The action of an event can have one of several forms of assignment, illustrated in Table 7.2. Here x is a variable, $E(t, v)$ is an expression, and $P(t, v, x)$ is a predicate. The first assignment form is deterministic. In the second row, the assignment is non-deterministic (for instance, assign a value within a non-empty set). The third row assigns a value to x according to the predicate defined, and it is also considered non-deterministic.

Three Possible Forms of an Event
E = begin S(v) end
E = when G(v) then S(v) end
E = any t when G(t,v) then S(t,v) end

Table 7.1. *Event forms*

Type	Generalized Substitution
Deterministic	x := E(t, v)
Non-deterministic	$x :\in E(t, v)$
Non-deterministic	$x :\mid P(t, v, x')$

Table 7.2. *Action forms*

7.1.3. *Refinement in Event-B*

In an Event-B development, rather than having a single large model, we are encouraged to construct the system in a series of successive layers, starting with an abstract representation of the system. The abstract model should provide a simple view of the system, focusing on the main purpose and key features of the system. The details of how the purpose is achieved are ignored in the abstraction. Details of functionality of the system are added gradually to the abstract model in a stepwise manner. This process is called refinement.

In Event-B modeling, we use proof to verify the consistency of a refinement. The definitions of some refinement proof obligations are described in section 7.1.4.

Refining an Event-B model can consist of context extension and machine refinement. Considering context extension, it is possible to add new sets, constants, and properties while retaining the old ones.

Refinement in Event-B has different views or classification. From Event-B notation point of view, refinement of a machine consists of:

1. Refining existing events:

(a) Adding new parameters, guards and actions to the existing abstract event: in this case the resulting concrete event is labeled as $extended$. In an $extended$ event, the existing parameters, guards and actions cannot be modified.

(b) Modifying parameters, guards, and actions of the existing abstract event: in this case, the resulting concrete event is labeled as $non - extended\,(refine)$. Adding new parameters, guards, and actions are allowed, too.

In both types, the guards of the concrete event must be proved to be stronger than its abstraction (guard strengthening).

2. Adding new events

The new event refines a dummy event in the abstraction, which does nothing (*skip*).

The new event does not diverge. It means that it should not take control forever. The new event can be labeled as:

- Convergent: Each convergent event requires a variant to ensure non-divergence.
- Anticipated: Events that will be introduced in a future refinement but are declared in anticipation.
- Ordinary: None of the others and the most commonly used.

3. Add new variables and invariants:

Introducing new variables usually results in (2) or (1.a) types of refinement. Sometimes, abstract variables can be replaced by new concrete variables. In this case, the refinement can result in (1.b). Variable replacement is called data refinement.

A gluing invariant connects the abstract variables to the concrete variables. In other words, it glues the state of the concrete model to that of its abstraction. The invariant of the concrete model including gluing invariants should be preserved for every event.

Each abstract event should be refined by at least one concrete event. One abstract event can be refined by more than one concrete event. This is called event splitting. Also, one concrete event can refine more than one abstract event. This is called event merging.

Refinement is the process of enriching or modifying the abstract model to introduce new functionality or add details of the current functionality. From another paint of view, there are two forms of refinement:

– Vertical Refinement (Structural Refinement): In this form, design details of current functionalities are added. This form of refinement may involve data refinement (3) and modifying abstract events (1.b). In the refinement level, the modified events are labeled as non-extended events.

– Horizontal Refinement (Superposition Refinement or Feature Augmentation): New functionalities of the system, which are not addressed in the abstract level, are introduced. Usually, it can be achieved by introducing new events (2), new variables

(3) or extending abstract events (1.a). In the refinement level, these concrete events are labeled as extended events.

7.1.4. *Proof obligations*

There are different proof obligations, which are generated by the Event-B tool, Rodin, during the development of a system using Event-B [ABR 08]. Here, we describe some of those that are most important. Considering Figure 7.4, machine $M2$ refines machine $M1$. Both of them see context Ctx. $M2$ contains two events, $evt3$ as a new event and $evt2$ as a refining event. It also contains some gluing invariants.

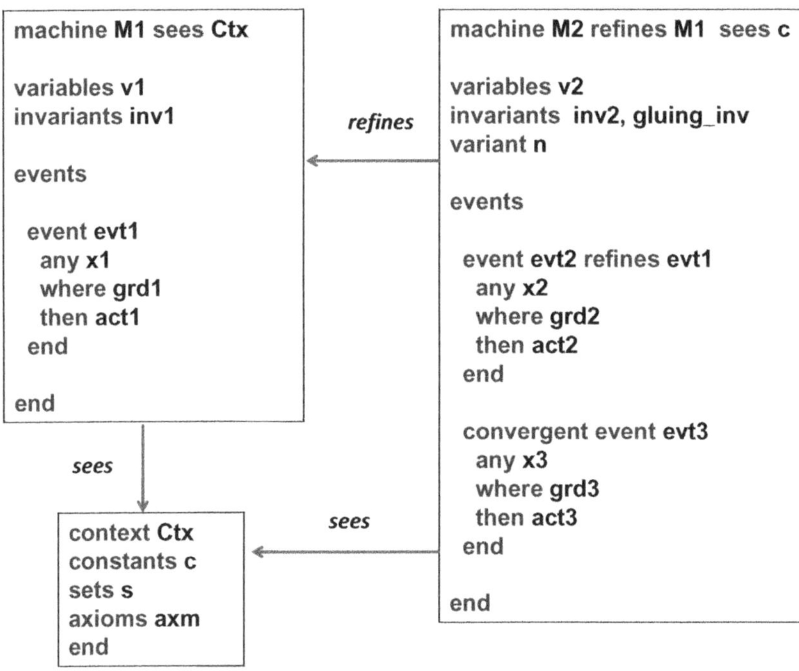

Figure 7.4. *An Event-B model (context Ctx, abstract machine $M1$, concrete machine $M2$)*

Table 7.3 contains a list of important proof obligation in Event-B modeling.

Here is an explanation for each of these proof obligations:

– *Well-definedness (WD)*: Ensure that an axiom, theorem, invariant, guard, action, variant is well-defined. When using the cardinality of a set, $card(S)$, it should be proved that the set, S, is finite.

– *Invariant Preservation (INT)*: Ensure that every invariant is preserved by each event. For instance, in Figure 7.4, one of the generated proof obligation's is $evt1$/$inv1$/INV, ensuring that $inv1$ is preserved by event $evt1$ in machine $M1$.

– *Feasibility (FIS)*: Ensure that each non-deterministic action is feasible. In Figure 7.4, for event $evt1$ in machine $M1$, this proof obligation is given: $evt1$ / $act1$ / FIS; this means there should exist values for variable $v1$ such that the assignment $act1$ is feasible.

– *Guard Strengthening (GRD)*: Ensure that each abstract guard is no stronger than the concrete ones in the refining event. As a result, when a concrete event is enabled, the corresponding abstract one is also enabled. For instance, for the model in Figure 7.4, $evt2$ / $grd1$ / GRD ensure that the abstract guard $grd1$ is weaker than the guards of the concrete event $evt2$.

– *Simulation (SIM)*: Ensure that each action in a concrete event simulates the corresponding abstract action. When a concrete event executes, the corresponding abstract event is not contradicted. In Figure 7.4 the simulation proof is $evt2$ / $act1$ / SIM.

– *Numeric Variant (NAT)*: Ensures that under the guards of each convergent event a proposed numeric variant is indeed a natural number. $evt3$ / NAT is the proof obligation generated for the model of Figure 7.4.

– *Decreasing of Variant (VAR)*: Ensures that each convergent event decreases the proposed numeric variant. As a consequence the new event does not take control forever. $evt3$ / VAR in Figure 7.4 ensures that event $evt3$ does not take control forever.

Well-definedness	x / WD	x is the name of axiom, theorem, invariant, guard, action, variant
Invariant preservation	evt / inv / INV	evt is the event name, inv is the invariant name
Feasibility of a non-deterministic event action	evt / act / FIS	evt is the event name, act is the action name
Guard strengthening	evt / grd / GRD	evt is the concrete event name, grd is the abstract guard name
Action simulation	evt / act / SIM	evt is the concrete event name, act is the abstract action name
Natural number for a numeric variant	evt / NAT	evt is the new event name
Decreasing of variant	evt / VAR	evt is the new event name

Table 7.3. *Proof obligations in Event-B*

7.1.5. *A comparison between Event-B and other formal methods*

Classical B, Z, and VDM have a one-to-one operation refinement, meaning that one abstract operation is refined by only one concrete operation. There is no facility for introducing new events in refinements in these formal methods. Event-B is more flexible as it bases its refinement on action systems. Also, event merging and event splitting are provided in Event-B refinement. Although Event-B is an extension of Classical B, there are some differences between them:

– The model structure is different. In Event-B, the context as the static part of the system and the machine as the dynamic part of the system are explicitly separated. In the B method, a machine contains both parts.

– In the B method, operations are called by other operations while in Event-B, the enabled events are continually executed in a non-deterministic manner. Since in Event-B, we are modeling reactive systems, the events are not called and the model controls its behavior by non-deterministically choosing the enabled events.

– A B method operation contains pre-conditions, which express formally what is to be proved when the operation is invoked. The caller of an operation is responsible for ensuring that pre-conditions of the called operation are satisfied before calling it. The called operation can assume that its pre-conditions are satisfied, and it does not need to check its pre-conditions. In contrast, an Event-B event contains guards. An event can be executed only when its guards hold. In Event-B, enabled events are non-deterministically chosen to execute.

– Refinement is more general in Event-B. Introducing new events is an important ability in Event-B refinement.

7.2. Rodin as an Event-B tool

Rodin [ABR 10, EVE] is an open source software tool for formal modeling and proving in Event-B. Rodin has an open platform and is an extensible and adaptable modeling tool. The ProB animator [WIK 02, LEU 08], UML-B [WIK 03, SNO 06], B2LaTeX [WIK 01] and model decomposition [SIL 11] are good examples of plug-in developments; ProB is a model checker, which checks the consistency of B machines; UML-B maps a graphical formal modeling notation to the Event-B language; B2LaTex is used for translating Event-B models into LaTeX documents; and model decomposition for decomposition of a model into sub-models. Decomposition will be explained in section 7.3.

Like programming tools, Rodin carries out many tasks automatically, and provides fast feedback in the case of changes in a model text. While a programming tool provides feedback to the programmer by compiling and executing a program, Rodin provides feedback to modellers by generating proof obligations and verifying these using automated provers.

Rodin is an integration between modeling and proving. As described in previous sections, proving is an essential part of modeling. The proof obligations define what is to be proved for an Event-B model. Discharging all proof obligations of a model shows that all model properties are consistent. Sometimes, a model can be changed using proof errors. When a proof obligation cannot be charged, it shows that there is an inconsistency in the model. This leads us to learn more about the system to change the model in an inconsistent way. Therefore, during modeling, we can learn about the system and eliminate misunderstandings. We can also learn new requirements by proving the failed proof obligations.

7.3. Event-B model decomposition

7.3.1. *Overview*

Model decomposition predated Event-B and is found in action systems [BAC 89]. In developing a model in Event-B, one of the key features is introducing new events and new state variables during refinement. As a consequence, it usually ends up with many events and many variables in the last refinement level. Dealing with a large number of events and variables can be complex; particularly when we need to refine just a few variables and events and so other variables and events play no role in the refinement.

Model decomposition in Event-B [SIL 11] is intended to decrease the complexity and increase the modularity of a large Event-B model, especially after several layers of refinements. The idea of model decomposition is cutting a huge model into smaller pieces called sub-models, which we can more easily deal with than the first model, and each of them can be refined separately.

Distribution of proof obligations into several sub-models is one of the major results of model decomposition, which is expected to be easier to discharge. The further refinements of independent sub-models in parallel is a benefit of model decomposition. Moreover, the possibility of team development after model decomposition seems useful in developing a big system.

An overview of the model decomposition in Event-B is illustrated in Figure 7.5. As presented, the model becomes bigger during refinement layers and with decomposition, it is split into smaller sub-models. Then, each sub-model can be refined independently.

7.3.2. *Decomposition styles*

There are two ways of decomposing an Event-B model, *shared variable* and *shared event* [SIL 11]. The shared event approach seems particularly suitable for message-passing distributed programs, whereas the shared variable approach seems more suitable for concurrent programs [BUT 97]. In shared event model decomposition, variables are partitioned among the sub-models, whereas in the shared variable approach, events are partitioned among the sub-models. Details are explained in the next section. A model decomposition plug-in [SIL 11] in the Rodin platform provides tool support for both styles of model decomposition.

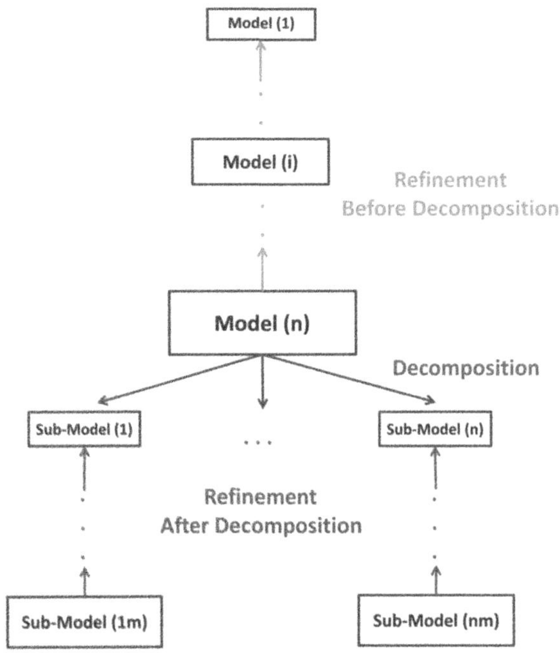

Figure 7.5. *Model decomposition in Event-B*

7.3.2.1. *Shared variable style*

Shared variable decomposition, illustrated in Figure 7.6, is proposed by Abrial and Hallerstede [ABR 07]. Machine *M* is decomposed into machine *M1* and *M2*. The solid lines show relationships between events and variables in each machine.

The shared variable decomposition does not permit event sharing and a variable can be split into different sub-models. This variable is called a shared variable. First, the events of *M* are partitioned among *M1* and *M2*. Then, the variables of *M* are distributed according to the event partition. Variables *v1* and *v3* are private variables since they are accessed by events of only one sub-model, *e1* in *M1* and *e4* in *M2*, respectively. Variable *v2* is a shared variable, which is accessed by event *e2* in *M1* and *e3* in *M2*. External event of *e2_ext* is built in *M2*, since *e2* modifies the shared variable *v2* in *M1*. The invariant distribution is done according to variable distribution. An invariant belongs to a sub-model if all variables used in that invariant belong to that sub-model.

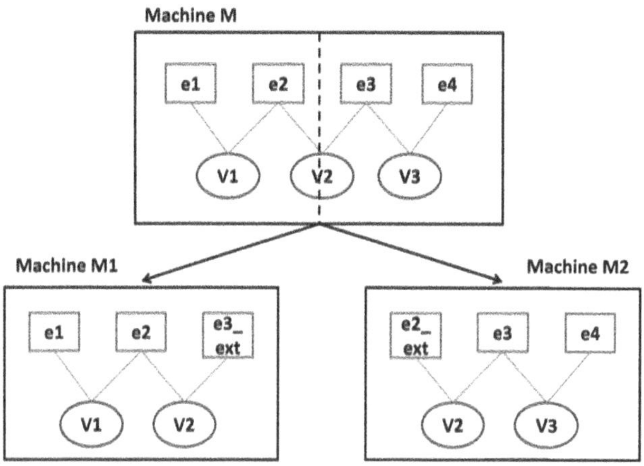

Figure 7.6. *Shared variable decomposition*

7.3.2.2. *Shared event style*

Figure 7.7 illustrates shared event decomposition proposed by Butler [BUT 09]. Variables of the machine *M* are partitioned among the sub-models, *M1* and *M2*. After the variable partition, it is necessary to split the events according to the variable partition. Events using variables allocated to different sub-models, *e2* using *v1* from *M1* and *v2* from *M2*, are called shared events and must be split. Part of the shared event, which is corresponding to each variable, *e2_1* and *e2_2*, is used to build sub-model events. Invariant distribution is similar to shared variable decomposition.

7.4. Case study: metro system

This section describes how the modeling, refinement, and decomposition techniques presented in the previous sections can be applied in practice. We aim to develop a system that becomes more complex with each refinement step, preserves its properties (requirements) and re-uses existing developments and proofs as much as possible. A safety-critical metro system case study is developed. This version is a simplified version of a real system but tackles points where the techniques outlined in the previous sections become relevant: stepwise incrementation of the complexity of the system being modeled, sub-components communication, stepwise addition of requirements at each refinement level, refinement of decomposed sub-components. Although this system is initially modeled as a single component, it can be seen as a distributed system where the initial model is split into smaller sub-components that communicate via shared events. The split is achieved through a shared event decomposition and the sub-components are further refined independently. After several refinements, we reach a refinement that fits an existing generic development for metro doors. Using that development as a pattern, two models are instantiated accordingly.

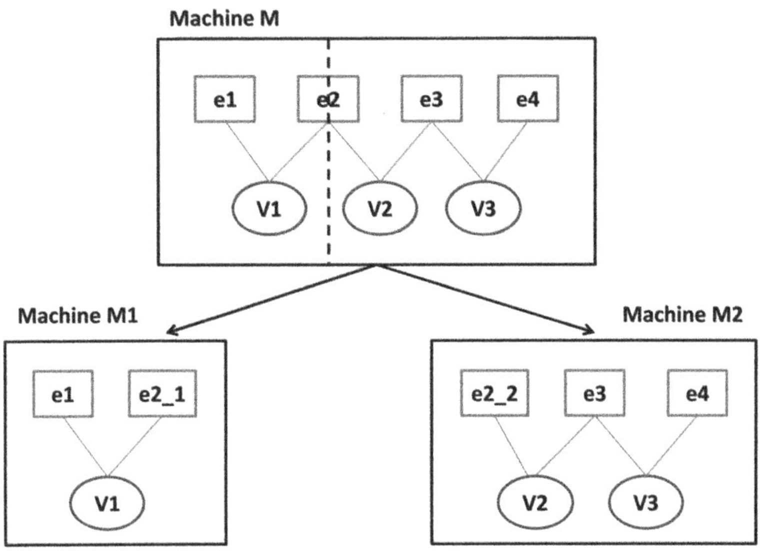

Figure 7.7. *Shared event decomposition*

7.4.1. *Overview of the safety-critical metro system*

The safety-critical metro system case study describes a formal approach for the development of embedded controllers for a metro system[1]. Butler [BUT 02] provides a description of embedded controllers for a railway using classical B. The railway system is based on the French train system. Our starting point is based on that work but applied to a metro system. That work goes as far as our first decomposition originating three sub-components. We augment that work by refining each sub-component, introducing further details, and more requirements to the model. Moreover, in the end, we instantiate emergency and service doors for the metro system.

The metro system is characterized by trains, tracks circuits (also called sections or CDV: *Circuit De Voie*, in French), and a communication entity that allows the interaction between trains and tracks. The trains circulate in sections and before a train

1 A version of this model is available online at http://eprints.ecs.soton.ac.uk/23135/.

enters or leaves a section, a permission notification must be received. In case of a hazardous situation, trains receive a notification to brake. The track is responsible for controlling the sections, changing switch directions (switch is a special track that can be divergent or convergent as seen in Figure 7.8), and sending signaling messages to the trains.

Figure 7.9 shows a schematic representation of the metro system decomposed into three sub-components. Initially, the metro system is modeled as a whole. Global properties are introduced and proved to be preserved throughout refinement steps. The abstract model is refined in three levels (*MetroSystem_M0* to *MetroSystem_M3*) before we apply the first decomposition. We follow a general top-down guideline to apply decomposition:

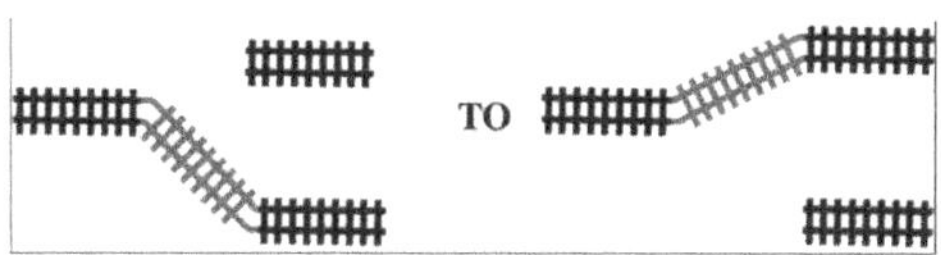

(a) Divergent Switch

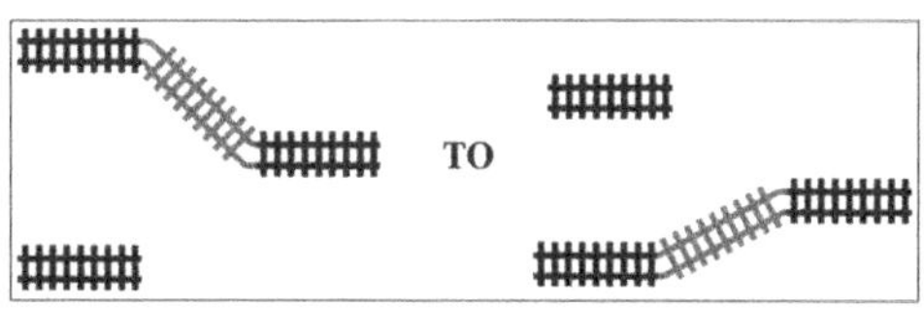

(b) Convergent Switch

Figure 7.8. *Different types of switches: divergent and convergent*

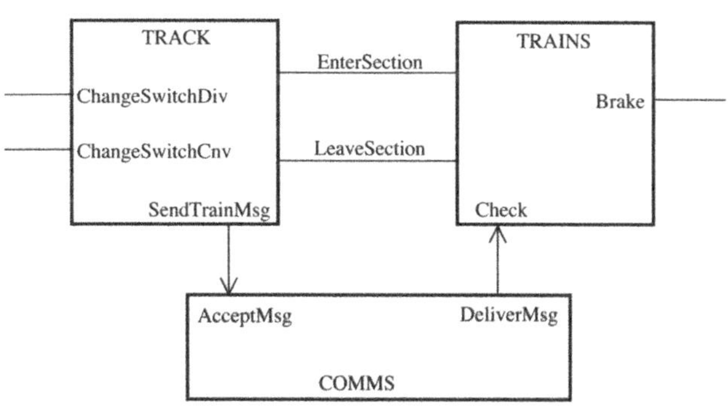

Figure 7.9. *Components of metro system*

Stage 1: Model system abstractly, expressing all the relevant global system properties.

Stage 2: Refine the abstract model to fit the decomposition (preparation step).

Stage 3: Apply decomposition.

Stage 4: Develop independently the decomposed parts.

For instance, **Stage 1** is expressed by refinements *MetroSystem_M0* to *MetroSystem_M3*. *MetroSystem_M3* is also used as the preparation step before the decomposition corresponding to **Stage 2**. The model is decomposed into three parts: *Track*, *Train*, and *Middleware* as described in **Stage 3**. This step allows further refinements of the individual sub-components corresponding to **Stage 4**. The following decompositions follow a similar pattern.

An overview of the development can be seen in Figure 7.10. After the first decomposition, sub-components can be further refined. Train global properties are introduced in *Train* leading to several refinements until *Train_M4* is reached. *Train_M4* is decomposed into *Leader Carriage* and *Carriage*. We are interested in refining the sub-component corresponding to carriages to introduce doors requirements. These requirements are extracted from real requirements for metro carriage doors.

7.4.2. *Abstract model: MetroSystem_M0*

We model a system constituted by trains that circulate on tracks. The tracks are divided into smaller parts called sections. The most important (safety) global property introduced at this stage states that two trains cannot be in the same section at the same time (which would mean that the trains might collide).

We need to ensure some properties regarding the routes (set of track sections):

– Route sections are all connected: sections should be all connected and cannot have empty spaces between them.

– There are no loops in the route sections: sections cannot be connected to themselves and cannot introduce loops.

These properties can be preserved if we represent the routes as a transitive closure relation. We use the no-loop property proposed by Abrial [ABR 08] and used to model a tree-structured file system in Event-B [DAM 08]: a context is defined and this property is proved over track section relations and functions. The reason we choose this formulation, instead of transitive closure, which is generally used, is to make the model easier to prove. Context $TransitiveClosureCtx$ containing the transitive closure property can be seen in Figure 7.11.

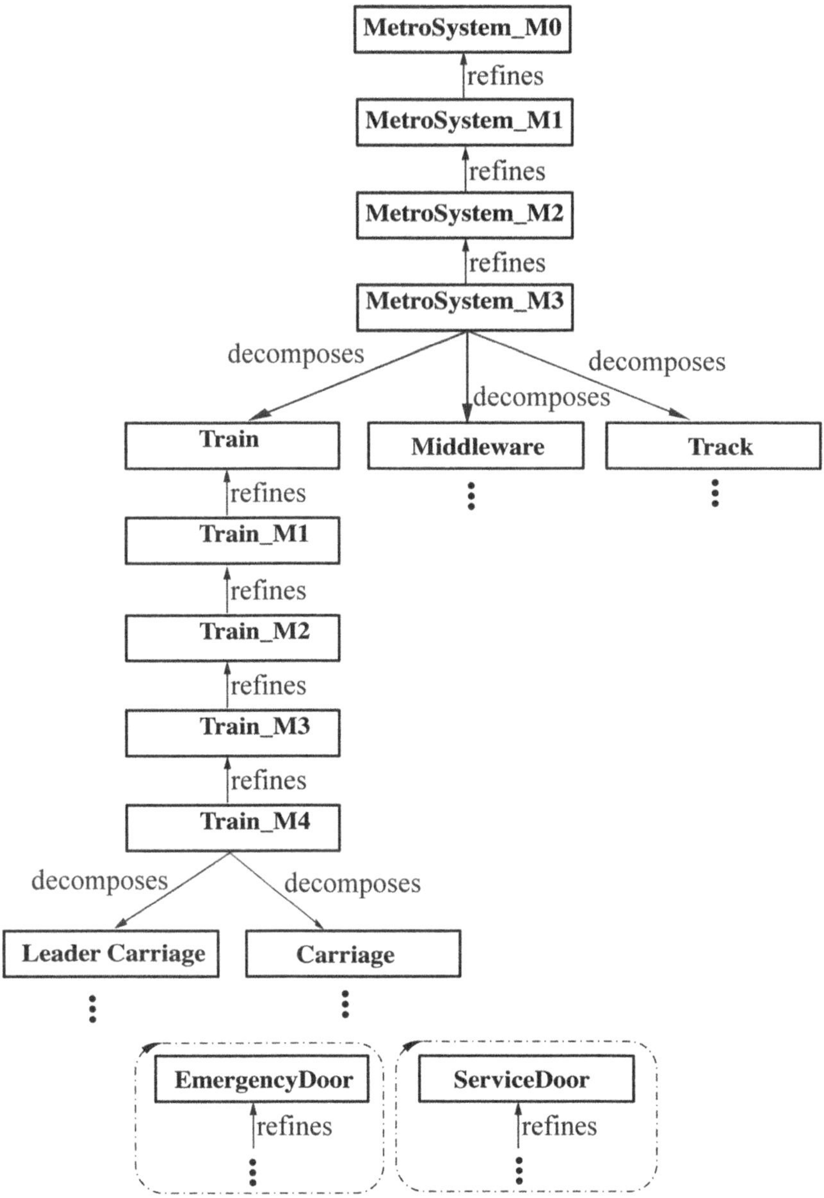

Figure 7.10. *Overall view of the safety-critical metro system development*

```
context TransitiveClosureCtx

constants cdvrel // type of relation on sections
          tcl // transitive closure of an cdvrel
          cdvfn // type of function on sections */

sets CDV // Track Sections

axioms
  @axm1 cdvrel = CDV ↔ CDV
  @axm2 cdvfn = CDV ⇸ CDV
  @axm3 tcl ∈ cdvrel → cdvrel
  @axm4 ∀r·(r∈cdvrel ⇒ r ⊆ tcl(r)) // r included in tcl(r)
  @axm5 ∀r·(r∈cdvrel ⇒r;tcl(r) ⊆ tcl(r)) // unfolding included in tcl(r)
  @axm6 ∀r,t·(r∈cdvrel ∧ r⊆t ∧ r;t⊆t ⇒ tcl(r)⊆t) // tcl(r) is least
  theorem @thm1 cdvfn ⊆ cdvrel
  theorem @thm2 ∀r·r∈cdvrel ⇒ tcl(r) = r ∪ (r;tcl(r)) // tcl(r) is a fixed
point
  theorem @thm3 ∀t·t∈cdvfn∧(∀s·s⊆t∼[s]⇒s=∅)⇒tcl(t)∩(CDV ◁ id)=∅
theorem @thm4 tcl(∅) = ∅
end
```

Figure 7.11. *Context TransitiveClosureCtx*

Set CDV represents all the track sections in our model. Constant tcl, which is a transitive closure, is defined as a total function mapped from $CDV \leftrightarrow CDV$ to $CDV \leftrightarrow CDV$. Giving $r \in CDV \leftrightarrow CDV$, the transitive closure of r is the least x satisfying $x = r \cup r; x$ [DAM 08]. Difficult transitive closure proofs in machines are avoided by using Theorems, such as theorem $thm3$ shown in Figure 7.11: for $s \subseteq CDV$ and t as a partial function $CDV \nrightarrow CDV$, $s \subseteq t^{-1}[s]$ means that s contains a loop in the t relationship. Hence, this states that the only such set that can exist is the empty set and thus the t structure cannot have loops. This theorem has been proved using the interactive prover of Rodin. The strategy to prove this theorem is to use proof by contradiction [DAM 08].

We define the environment of the case study (static part) with context *MetroSystem_C0* that extends *TransitiveClosureCtx* as seen in Figure 7.12. Set $TRAIN$ represent all the trains in our model. Several track properties are described in the axioms:

– The constant net represents the total possible connectivity of sections (all possible routes subject to the switches positions) defined as relation $CDV \leftrightarrow CDV$ ($axm1$). No circularity is allowed as described by $axm2$. Moreover, the no loop property for net is expressed by axiom $axm11$. Theorems $thm1$ state that net preserves transitive closure.

– Switches (*aiguillages* in French) are sections ($axm3$) that cannot be connected to each other ($axm6$). They are represented by aig_cdv divided into two kinds: div_aig_cdv for divergence switches and cnv_aig_cdv for convergent switches. Moreover, switches have at most two predecessors and one successor or one predecessor and two successors ($axm10$).

– Non-switches have at most one successor and at most one predecessor ($axm9$).

```
context MetroSystem_C0 extends TransitiveClosureCtx

constants aig_cdv // Switches
          net // Total connectivity of sections */
          div_aig_cdv // divergent switches 1->2
          cnv_aig_cdv // convergent switches 2->1
          next0

sets TRAIN

axioms
  @axm1 net ∈ CDV ↔ CDV // net represents the connectivity between track sections /*
  @axm2 net ∩(CDV ◁ id)=∅ // no cdv is connected to itself
  @axm3 aig_cdv ⊆ CDV // aig_cdv is a subset of CDV representing cdv that are switches
  @axm4 div_aig_cdv  ⊆ aig_cdv // div_aig_cdv ⊆ aig_cdv
  @axm5 cnv_aig_cdv ⊆ aig_cdv
  @axm6 div_aig_cdv ∩ cnv_aig_cdv = ∅
  @axm7 finite(net) // explicite declaration to simplify the proving
  @axm8 (aig_cdv × aig_cdv) ∩ net = ∅ // switches are not directly connected
  @axm9 ∀cc·(cc ∈ (CDV\aig_cdv) ⇒ card(net[{cc}]) ≤1 ∧ card(net~[{cc}])≤1) // non
switch cdv has at most one successor and at most one predecessor
  @axm10 ∀cc·( cc ∈ aig_cdv ⇒ ( (card(net[{cc}])≤2 ∧ card(net~[{cc}])≤1) ∨ (
card(net[{cc}]) ≤1 ∧ card( net~[{cc}])≤2 ))) // switch cdv has at most two predecessors
and one successor or one predecessor and two successors
  @axm11 tcl(net)∩id=∅ // No-loop property
  theorem @thm1 tcl(net) = net ∪ (net;tcl(net))// the transitive closure of net is
equal to net ∪ net;tcl(net)
end
```

Figure 7.12. *Context MetroSystem_C0*

In addition to the global property defined by invariant $inv13$ in Figure 7.13a, the following system properties are added to the Event-B model:

1. The trains (variable $trns$) circulate on tracks. The current route based on current positions of switches is defined by $next$: a partial injection $CDV \rightarrowtail\!\!\!\!\!\!\!+\; CDV$. $next$ is a subset of net ($inv1$) preserving the transitive closure property as described by theorem $thm1$, $thm2$ and does not have loops ($thm3$). Sections occupied by trains are represented by variable $occp$. These sections also preserve the transitive closure property as seen by $thm4$;

2. A train occupies at least one section and the section corresponding to the beginning and end of the train is represented by variables $occpA$ and $occpZ$, respectively. Note that $next$ does not indicate the direction that a train is moving in: the direction can be $occpA$ to $occpZ$ or $occpZ$ to $occpA$. These two variables point to the same section if the train only occupies one section ($inv11$).

The system proceeds as follows: trains modeled in the system circulate by entering and leaving sections (events $enterCDV$ and $leaveCDV$ in Figure 7.13b), ensuring that the next section is not occupied ($grd9$ in $enterCDV$) and updating all the sections occupied by the train ($act1$ and $act2$ in both events). At this abstract level, event $modifyTrain$ modifies a train defining the set of occupied sections for a train t. A train changes speed, brakes, or stops braking in events $changeSpeed$, $brake$, and $stopBraking$. When event $brake$ occurs, train t is added to a set of braking trains (variable $braking$). Variable $next$ represents the current connectivity of the trail based on the positions of switches. The current connectivity can be updated by changing

```
machine MetroSystem_M0 sees MetroSystem_C0

variables next // Current connectivity based on switch positions
          trns // Set of trains on network
          occp // Occupancy function for section
          occpA // Initial cdv occupied by train
          occpZ // Final   cdv occupied by train
          braking speed

invariants
  @inv1 next ⊆ net
  @inv2 next ∈ CDV ⤔ CDV
  @inv3 trns ⊆ TRAIN
  @inv4 occp ∈ CDV ↔ trns
  @inv5 occpA ∈ trns → CDV
  @inv6 ∀tt·(tt∈trns ⇒ occpA(tt) ∈ occp∼[{tt}])
  @inv7 occpZ ∈ trns → CDV
  @inv8 ∀tt·(tt∈trns ⇒ occpZ(tt) ∈ occp∼[{tt}])
  @inv9 braking ⊆ trns
  @inv10 speed ∈ trns → ℕ
  @inv11 ∀tt·tt∈trns ∧ card(occp∼[{tt}])>1 ⇒ occpA(tt) ≠ occpZ(tt)
  @inv12 finite(occp∼)
  @inv13 ∀t1,t2·t1∈trns ∧ t2∈trns ∧ t1≠t2 ⇒ occp∼[{t1}]∩occp∼[{t2}]=∅
   theorem @thm1 next ∈ cdvfn
  theorem @thm2 tcl(next) = next ∪ (next;tcl(next)) // tcl(next) is a fixed
point
  theorem @thm3 (∀s·s⊆next∼[s]⇒s=∅)⇒tcl(next)∩(CDV ◁ id)=∅ // next has no
loops
  theorem @thm4 ∀tt,s·tt∈trns ∧ s ⊆ next▷occp∼[{tt}] ⇒ tcl(s) = s ∪
(s;tcl(s))
```

(a) Variables, invariants in *MetroSystem_M0*

```
event enterCDV
  any t1 c1 c2
  where
    @grd1 t1 ∈ trns
    @grd2 c1 ∈ CDV
    @grd3 c2 ∈ CDV
    @grd4 speed(t1)>0
    @grd5 c1 = occpZ(t1)
    @grd6 c1∈dom(next)
    @grd7 c2 = next(occpZ(t1))
    @grd8 ∀tt·tt∈trns ∧ card((occp ∪ {c2 ↦ t1})∼[{tt}])>1
         ⇒ (occpZ◁{t1 ↦ c2})(tt) ≠ occpA(tt)
    @grd9 c2 ∉ dom(occp)
  then
    @act1 occpZ(t1) ≔ c2
    @act2 occp≔occp ∪ { c2 ↦ t1}
end

event leaveCDV
  any t1 c1 c2
  where
    @grd1 t1 ∈ trns
    @grd2 c1 ∈ CDV
    @grd3 c2 ∈ CDV
    @grd4 speed(t1)>0
    @grd5 c1∈dom(next)
    @grd6 c1=occpA(t1)
    @grd7 c2=next(c1)
    @grd8 occpA(t1)≠occpZ(t1)
    @grd9 c2 ∈ (occp\{c1↦t1})∼[{t1}]
    @grd10 ∀tt·tt∈trns ∧ card(((occp \ {c1 ↦ t1}))∼[{tt}])>1
          ⇒ (occpA◁{t1 ↦ c2})(tt)≠occpZ(tt)
  then
    @act1 occpA(t1)≔c2
    @act2 occp ≔ occp\{c1↦t1}
end

event changeSpeed
  any t1 s1
  where
    @grd1 t1 ∈ trns
    @grd2 s1 ∈ ℕ
    @grd3 t1∈ braking ⇒ s1<speed(t1)
  then
    @act1 speed(t1) ≔ s1
end

event brake
  any t1
  where
    @grd1 t1 ∈ TRAIN
    @grd2 t1∈trns\braking
  then
    @act1 braking≔braking∪ {t1}
end

event stopBraking
  any t1
  where
    @grd1 t1 ∈ TRAIN
    @grd2 t1∈braking
  then
    @act1 braking≔braking\{t1}
end

event switchChangeDiv
  any ac c1 c2
  where
    @grd1 ac ∈ div_aig_cdv
    @grd2 c1 ∈ CDV
    @grd3 c2 ∈ CDV
    @grd8 c2 ∉ ran (next)
    @grd4 (ac ↦ c1) ∈ next
    @grd5 (ac ↦ c2) ∈ net
    @grd6 c1 ≠ c2
    @grd7 ac ∉ dom(occp)
  then
    @act1 next ≔ next  {ac ↦ c2}
end

event switchChangeCnv
  any ac c1 c2
  where
    @grd1 ac ∈ cnv_aig_cdv
    @grd2 c1 ∈ CDV
    @grd3 c2 ∈ CDV
    @grd8 c2 ∉ dom (next)
    @grd4 (c1 ↦ ac) ∈ next
    @grd5 (c2 ↦ ac) ∈ net
    @grd6 c1 ≠ c2
    @grd7 ac ∉ dom (occp)
  then
    @act1 next ≔ ({c1}⩤next) ∪ {c2 ↦ ac}
end

event addTrain
  any t oc
  where
    @grd1 t ∈ TRAIN\trns
    @grd2 oc ∈ CDV
    @grd3 oc ∉ dom(occp)
  then
    @act1 trns≔trns ∪{t}
    @act2 speed(t)≔0
    @act3 occpA(t) ≔ oc
    @act4 occpZ(t) ≔ oc
    @act5 occp ≔ occp ∪ {oc↦t}
end

event modifyTrain
  any t ocA oc
  where
    @grd1 ocA∈dom(next)
    @grd2 t ∈ trns
    @grd3 oc ⊆ CDV
    @grd4 ocA ∈ oc
    @grd5 oc ∩ dom(occp)=∅
    @grd6 finite(oc)
    @grd7 occpZ(t)∈dom(next)
    @grd8 card(oc)=0 ⇒ocA = occpZ(t)
    @grd9 card(oc)≥1
          ⇒ occpZ(t) ≠ ocA ∧ next(occpZ(t))∈oc
    @grd10 next(ocA)∉oc
  then
    @act1 occpA(t) ≔ ocA
    @act2 occp ≔ occp ∪ (oc×{t})
end
```

(b) Events of *MetroSystem_M0*

Figure 7.13. *Variables, invariant and events of MetroSystem_M0*

convergent/divergent switches in events $switchChangeDiv$ and $switchChangeCnv$ as seen in Figure 7.13b.

7.4.3. *First refinement: MetroSystem_M1*

MetroSystem_M1 refines *MetroSystem_M0*, incorporating the communication layer and an emergency button for each train. The communication works as follows: a message is sent from the tracks, stored in a buffer, and read in the recipient train. The properties to be preserved for this refinement are as follows:

1. Messages are exchanged between trains and tracks. If a train intends to move to an occupied section, the track sends a message negating the access to that section and the train should brake.

2. As part of the safety requirements, all trains have an emergency button.

3. While the emergency button is enabled, the train continues braking and cannot speed up.

Now, the system proceeds as follows: trains that enter and leave sections must take into account the messages sent by the tracks. Therefore, events corresponding to entering and leaving the section need to be strengthened to preserve this property. The requirement concerning the space required for the train to halt is a simplification of a real metro system and could require adjustments to replicate the real behavior (for instance the occupied sections of a train could be defined as the sum of the sections directly occupied by the train and the sections indirectly occupied by the same train that correspond to the sections required for the train to halt). Nevertheless, in real systems, trains can have a built-in way to detect the required space to break. For instance, in Communication-Based Train Control (CBTC [TSD 12, FAL 09]) systems, this is called the *stopping distance downstream*.

The messages are represented by variables $tmsgs$ that store the messages (buffer) sent from the tracks and $permits$ that receive the message in the train, expressing property 1. At this level, the messages are just Boolean values assessing whether a train can move to the following section (check if the section is free): if TRUE the train can move; if FALSE the next section is occupied and the train should brake. New event $sendTrainMsg$ models the message sending. The reception of messages is modeled in event $recvTrainMsg$ where the message is stored in $permit$ before $tmsgs$ is reset. The guards of event $brake$ are strengthened to allow a train to brake when $permit(t) =$ FALSE or when the emergency button is activated (guard $grd3$ in Figure 7.14b). Property 2 is expressed by adding variable $emergency\_button$. The activation/deactivation of the emergency button occurs in the new event $toggleEmergencyButton$. Property 3 is expressed by guard $grd3$ in the event $stop$ $Braking$: a train can only stop braking if the emergency button is not enabled.

```
machine MetroSystem_M1 refines MetroSystem_M0  sees MetroSystem_C0

variables next trns occp occpA occpZ
          braking speed
          tmsgs permit emergency_button

invariants
  @inv1 tmsgs ∈ trns → ℙ(BOOL)
  @inv2 permit ∈ trns → BOOL
  @inv3 emergency_button ∈ trns → BOOL
```

(a) Variables and invariants in *MetroSystem_M1*

```
event brake refines brake
  any t1
  where
    @grd1 t1 ∈ TRAIN
    @grd2 t1∈trns\braking
    @grd3 permit(t1) = FALSE
          ∨ emergency_button(t1)=TRUE
  then
    @act1 braking≔braking ∪ {t1}
end

event sendTrainMsg
  any t1
  where
    @grd1 t1 ∈ trns
    @grd2 tmsgs(t1) = ∅
  then
    @act1 tmsgs(t1)≔ {bool(
          occpZ(t1)∈dom(next)
          ∧next(occpZ(t1)) ∉ dom(occp))}
end

event stopBraking refines stopBraking
  any t1
  where
    @grd1 t1 ∈ TRAIN
    @grd2 t1∈braking
    @grd3 emergency_button(t1) = FALSE
  then
    @act1 braking≔braking\{t1}
end

event recvTrainMsg
  any t1 bb
  where
    @grd1 t1 ∈ trns
    @grd2 bb ∈ tmsgs(t1)
  then
    @act1 permit(t1) ≔ bb
    @act2 tmsgs(t1) ≔ ∅
end

event toggleEmergencyButton
  any t value
  where
    @guard t ∈ trns
    @guard1 value ∈ BOOL
  then
    @act1 emergency_button(t)≔ value
end
```

(b) Some events of *MetroSystem_M1*

Figure 7.14. *Excerpt of MetroSystem_M1*

7.4.4. *Second refinement: MetroSystem_M2*

In this refinement, we introduce train doors and platforms where the trains can stop to load/unload. When stopped, a train can open its doors. The properties to be preserved are as follows:

1. If a train door is opened, then the train is stopped. In contrast, if the train is moving, then its doors are closed.

2. If a train door is opened, that either means that the train is on a platform or there was an emergency and the train had to stop suddenly.

3. A train door cannot be allocated to different trains.

We consider that platforms are represented by single sections. A train is on a platform if one of the occupied sections corresponds to a platform. Doors are introduced as illustrated in Figure 7.15a by sets $DOOR$ and their states are represented by $DOOR\_STATE$. Variables $door$ and $door\_state$ represent the train doors and their current states as seen in Figure 7.15b: all trains have allocated a subset of doors ($inv2$). Several invariants are introduced to preserve the required properties: property 1 is defined by invariants $inv4$ and $inv5$; property 2 is defined by invariant $inv7$; property 3 is stated by $inv3$; theorem $thm1$ is used for proving purposes (if no doors are open, then all doors are closed).

```
context MetroSystem_C1 extends MetroSystem_C0

constants OPEN CLOSED PLATFORM

sets DOOR_STATE DOOR

axioms
  @axm1 partition(DOOR_STATE, {OPEN}, {CLOSED})
  @axm2 PLATFORM ⊆ CDV
end
```

(a) Context *MetroSystem_C1*

```
machine MetroSystem_M2 refines MetroSystem_M1  sees MetroSystem_C1

variables next trns occp occpA occpZ
          braking speed tmsgs permit
          door door_state emergency_button

invariants
  @inv1 door_state ∈ DOOR → DOOR_STATE
  @inv2 door ∈ trns → ℙ(DOOR)
  @inv3 ∀t1,t2·t1 ∈ dom(door) ∧ t2 ∈ dom(door) ∧ t1 ≠t2
        ⇒ door(t1) ∩ door(t2) = ∅
  @inv4 ∀t·t ∈ dom(door) ⇒(∃d·d⊆door(t) ∧ door_state[d]={OPEN}
        ⇒ speed(t)=0)
  @inv5 ∀t·t ∈ dom(door) ∧ speed(t) > 0
        ⇒ door(t) ⊆ door_state∼[{CLOSED}]
  @inv6 ∀t,d·t ∈ dom(door) ∧ d ∈ door(t) ∧ PLATFORM ∩ occp∼[{t}]≠∅
        ⇒ door_state(d) ∈ {OPEN, CLOSED}
  @inv7 ∀t·t ∈ dom(door) ∧ door(t) ∩ door_state∼[{OPEN}] ≠ ∅
        ⇒ PLATFORM ∩ occp∼[{t}]≠∅ ∨ emergency_button(t) = TRUE
  theorem @thm1 ∀t·t ∈ dom(door) ∧ door(t) ∩ door_state∼[{OPEN}] =∅
          ⇒ door(t)⊆door_state∼[{CLOSED}]
```

(b) Variables, invariants in *MetroSystem_M2*

```
event toggleEmergencyButton
refines toggleEmergencyButton
  any t value
  where
    @grd1 t ∈ dom(door)
    @grd2 value ∈ BOOL
    @grd3 door(t) ∩ door_state∼[{OPEN}] ≠ ∅
          ∧ PLATFORM ∩ occp∼[{t}]=∅
          ⇒ value = TRUE
  then
    @act1 emergency_button(t)≔ value
end

event openDoor
  any t ds
  where
    @grd1 t ∈ dom(door)
    @grd2 speed(t) = 0
    @grd3 occp∼[{t}] ∩ PLATFORM ≠ ∅
          ∨ emergency_button(t) = TRUE
    @grd4 ds ⊆ door(t)
    @grd5 ∃d·d∈ds⇒door_state(d)=CLOSED
    @grd6 ds≠∅
  then
    @act1 door_state≔ door_state  (ds×{OPEN})
end

event closeDoor
  any t ds
  where
    @grd1 t ∈ dom(door)
    @grd2 speed(t) = 0
    @grd3 ds ⊆ door(t)
    @grd4 door_state[ds]={OPEN}
    @grd5 ds≠∅
  then
    @act1 door_state≔ door_state  (ds×{CLOSED})
end

event addDoorTrain
  any t d
  where
    @grd1 t ∈ trns
    @grd2 d ⊆ DOOR
    @grd3 ∀tr·tr∈dom(door) ∧ tr≠t
          ∧ door(tr)≠∅⇒d∩door(tr)=∅
    @grd5 speed(t)=0
    @grd7 d∩door(t)=∅
  then
    @act1 door(t)≔door(t)∪d
    @act2 door_state≔
            door_state(d×{CLOSED})
end

event removeDoorTrain
  any t d
  where
    @grd1 t ∈ dom(door)
    @grd2 d ⊆ DOOR
    @grd3 d ⊆ door(t)
    @grd4 door_state[d]={CLOSED}
    @grd5 speed(t)=0
  then
    @act1 door(t) ≔ door(t)\d
end

event leaveCDV refines leaveCDV
  any t1 c1 c2
  where
    @grd1 t1 ∈ dom(door)
    @grd2 c1 ∈ CDV
    @grd3 c2 ∈ CDV
    @grd4 speed(t1)>0
    @grd5 c1∈dom(next)
    @grd6 c1=occpA(t1)
    @grd7 c2=next(c1)
    @grd8 occpA(t1)≠occpZ(t1)
    @grd9 c2 ∈ (occp\{c1↦t1})∼[{t1}]
    @grd10 ∀tt·tt∈trns
    ∧ card(((occp \ {c1 ↦ t1}))∼[{tt}])>1
    ⇒ (occpA{t1 ↦ c2})(tt)≠occpZ(tt)
    @grd11 door(t1)∩door_state∼[{OPEN}]=∅
    @grd12 permit(t1)=TRUE
  then
    @act1 occpA(t1)≔c2
    @act2 occp ≔ (occp\{c1↦t1})
end

event changeSpeed refines changeSpeed
  any t1 s1
  where
    @grd1 t1 ∈ dom(door)
    @grd2 s1 ∈ ℕ
    @grd3 t1∈ braking ⇒ s1<speed(t1)
    @grd4 door(t1)∩door_state∼[{OPEN}]=∅
  then
    @act1 speed(t1) ≔ s1
end
```

(c) Some events of *MetroSystem_M2*

Figure 7.15. *Excerpt of MetroSystem_M2*

To preserve $inv5$, the guards of $changeSpeed$ (in Figure 7.14b) are strengthened by $grd4$ ensuring that whilst the train is moving, the train doors are closed. Also, events that model entering and leaving sections are affected, with the introduction of a similar guard ($grd11$ in $leaveCDV$). Adding/removing train doors is modeled in events $addDoorTrain$ and $removeDoorTrain$, respectively: to add/remove a door, the respective train must be stopped. If the train is stopped and either one of the occupied sections corresponds to a platform or the emergency button is activated (guard $grd3$), doors can be opened as seen in event $openDoor$. For safety reasons, event $toggleEmergencyButton$ is strengthened by guard $grd3$ to activate the emergency button whenever doors are open and the train is not on a platform.

7.4.5. *Third refinement and first decomposition: MetroSystem_M3*

This refinement does not introduce new details to the model. It corresponds to the preparation step before the decomposition. We want to implement a three-way shared event decomposition and therefore we need to separate the variables that will be allocated to each sub-component. In particular, for exchanged messages between the sub-components, the protocol will work as follows: messages are sent from *Track* and stored in the *Middleware*. After receiving the message, the *Middleware* forwards it to the corresponding *Train*. *Train* reads the message and processes it according to the content. This protocol allows a separation between *Train* and *Track* with the *Middleware* working as a bridge between these two sub-components.

The decomposition follows the steps described in section 7.3.2.2. Variables are distributed according to Figure 7.16. To avoid constraints during the decomposition process, predicates and assignments containing variables that belong to different sub-components are re-arranged in this refinement step.

Some guards need to be rewritten in the refined events. For instance, guard $grd10$ in event $leaveCDV$ needs to be rewritten so as not to include both variables $trns$ (sub-component $Train$) and $occp$ (sub-component $Track$). Therefore, it is changed from:

$$\forall tt \cdot tt \in \mathbf{trns} \wedge card((occp \cup \{c2 \mapsto t1\})^{-1}[\{tt\}]) > 1 \Rightarrow (occpZ \Leftarrow \{t1 \mapsto c2\})(tt) \neq occpA(tt)$$

to:

$$\forall tt \cdot tt \in \mathbf{dom(occpZ)} \wedge card((occp \cup \{c2 \mapsto t1\})^{-1}[\{tt\}]) > 1 \Rightarrow (occpZ \Leftarrow \{t1 \mapsto c2\})(tt) \neq occpA(tt)$$

(Figure 7.17).

Both predicates represent the same property since $trns$ corresponds to the domain of variable $occpZ$ (see $inv7$ in Figure 7.13a). In Figure 7.17, the original guard $grd3$ in $toggleEmergencyButton$ is rewritten to separate variables $occp$ and $door$.

In this case, an additional parameter $occpTrns$ representing the variable $occp$ is added ($grd4$). This additional parameter will represent the value passing between the resulting decomposed events: parameter $occpTrns$ is written the value of $occp$ and afterward it is read in guard $grd3$. Similarly, guard $grd4$ in event $openDoor$ must not include variables $occp$ and $emergency\_button$ and consequently parameter $occpTrns$ is added.

Sub-components $Train$, $Track$, and $Middleware$ are described in the following sections.

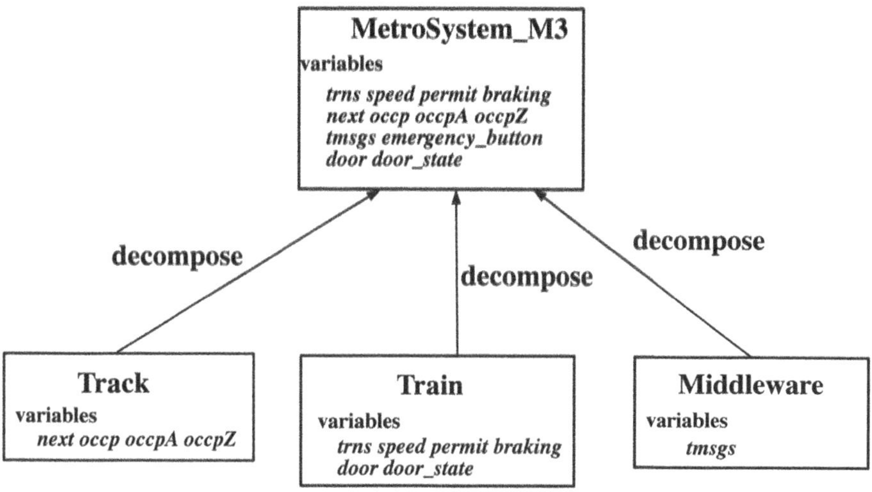

Figure 7.16. *MetroSystem_M3 (shared event) decomposed into Track, Train and Middleware*

7.4.6. *Machine Track*

Machine *Track* contains the properties concerning the sections in the metro system. Events corresponding to entering, leaving tracks, and changing switch positions are part of this sub-component resulting from the variables allocation for this sub-component: $next$, *occp*, *occpA* and $occpZ$. Event *sendTrainMsg* is also added since the messages are sent from the tracks as seen in Figure 7.18. The original events $toggleEmergency Button$ and $openDoor$ require $occp$ in their guards. Consequently, a part of these original events are included in this sub-component.

Note that the invariants defining the variables may change: in *MetroSystem_M1* variable *occp* is defined as $occp \in CDV \leftrightarrow trns$ (*inv4* in Figure 7.13a); in *Track* is $occp \in CDV \leftrightarrow TRAIN$ (which is the same as theorem $typing\_occp$: $occp \in \mathbb{P}(CDV \times TRAIN)$ in Figure 7.18). This is a consequence of the variable partition

```
event toggleEmergencyButton
refines toggleEmergencyButton
    any t value occpTrns
    where
      @grd1 t ∈ dom(door)
      @grd2 value ∈ BOOL
      @grd3 door(t) ∩ door_state~[{OPEN}] ≠ ∅
            ∧ PLATFORM ∩ occpTrns=∅
            ⇒ value = TRUE
      @grd4 occpTrns = occp~[{t}]
    then
      @act1 emergency_button(t)≔ value
  end

event openDoor refines openDoor
    any t occpTrns ds
    where
      @grd1 t ∈ dom(door)
      @grd2 speed(t) = 0
      @grd3 occpTrns = occp~[{t}]
      @grd4 occpTrns ∩ PLATFORM ≠ ∅
            ∨ emergency_button(t) = TRUE
      @grd5 ds ⊆ door(t)
      @grd6 ∃d· d∈ds⇒door_state(d)=CLOSED
      @grd7 ds≠∅
    then
      @act1 door_state≔ door_state  (ds×{OPEN})
  end

event leaveCDV refines leaveCDV
  any t1 c1 c2
  where
    @grd1 t1 ∈ dom(door)
    @grd2 c1 ∈ CDV
    @grd3 c2 ∈ CDV
    @grd4 speed(t1)>0
    @grd5 c1∈dom(next)
    @grd6 c1=occpA(t1)
    @grd7 c2=next(c1)
    @grd8 occpA(t1)≠occpZ(t1)
    @grd9 c2 ∈ (occp\{c1↦t1})~[{t1}]
    @grd10 ∀tt· tt∈dom(occpZ)
           ∧ card(((occp \ {c1 ↦ t1}))~[{tt}])>1
           ⇒ (occpA{t1 ↦ c2})(tt)≠occpZ(tt)
    @grd11 door(t1)∩door_state~[{OPEN}]=∅
    @grd13 permit(t1)=TRUE
  then
    @act1 occpA(t1)≔c2
    @act2 occp ≔ (occp\{c1↦t1})
end
```

Figure 7.17. *Preparation step before decomposition of MetroSystem_M3*

```
machine Track sees MetroSystem_C1

variables next occp occpA occpZ

invariants
  theorem @typing_occpZ occpZ ∈ ℙ(TRAIN × CDV)
  theorem @typing_occp occp ∈ ℙ(CDV × TRAIN)
  theorem @typing_next next ∈ ℙ(CDV × CDV)
  theorem @typing_occpA occpA ∈ ℙ(TRAIN × CDV)
  @MetroSystem_M0_inv1 next ⊆ net
  @MetroSystem_M0_inv2 next ∈ CDV ⤔ CDV
  @MetroSystem_M0_inv12 finite(occp~)

  event sendTrainMsg
    any t1 bb
    where
      @typing_t1 t1 ∈ TRAIN
      @typing_bb bb ∈ BOOL
      @grd3 bb = bool (occpZ(t1)∈dom(next)
            ∧ next(occpZ(t1))∉dom(occp) )
  end

   event enterCDV
     any t1 c1 c2
     where
       @typing_t1 t1 ∈ TRAIN
       @grd2 c1 ∈ CDV
       @grd3 c2 ∈ CDV
       @grd5 c1 = occpZ(t1)
       @grd6 c1∈dom(next)
       @grd7 c2 = next(occpZ(t1))
       @grd8 ∀tt· tt∈dom(occpZ)
             ∧ card((occp ∪ {c2 ↦ t1})~[{tt}])>1
             ⇒ (occpZ{t1 ↦ c2})(tt) ≠ occpA(tt)
       @grd9 c2∉dom(occp)
     then
       @act1 occpZ(t1) ≔ c2
       @act2 occp≔occp ∪ { c2 ↦ t1}
   end

  event openDoor
    any t occpTrns ds
    where
      @typing_t t ∈ TRAIN
      @typing_occpTrns occpTrns ∈ ℙ(CDV)
      @typing_ds ds ∈ ℙ(DOOR)
      @grd3 occpTrns = occp~[{t}]
      @grd7 ds≠∅
  end
  event leaveCDV
   any t1 c1 c2
   where
     @typing_t1 t1 ∈ TRAIN
     @grd2 c1 ∈ CDV
     @grd3 c2 ∈ CDV
     @grd5 c1∈dom(next)
     @grd6 c1=occpA(t1)
     @grd7 c2=next(c1)
     @grd8 occpA(t1)≠occpZ(t1)
     @grd9 c2 ∈ (occp\{c1↦t1})~[{t1}]
     @grd10 ∀tt· tt∈dom(occpZ)
            ∧ card(((occp \ {c1 ↦ t1}))~[{tt}])>1
            ⇒ (occpA{t1 ↦ c2})(tt)≠occpZ(tt)
   then
     @act1 occpA(t1)≔c2
     @act2 occp ≔ (occp\{c1↦t1})
  end
event toggleEmergencyButton
  any t value occpTrns
  where
    @typing_t t ∈ TRAIN
    @typing_occpTrns occpTrns ∈ ℙ(CDV)
    @grd2 value ∈ BOOL
    @grd4 occpTrns = occp~[{t}]
end
```

Figure 7.18. *Excerpt of Track*

since *trns* is not part of *Track* and therefore, the *occp* relation is updated with *trns*'s type: $TRAIN$ (see $inv3$ in Figure 7.13a). Variables *occpA* and *occpZ* are subject to the same procedure where the original invariant is a total function $trns \to CDV$ and in the sub-component, both become $\mathbb{P}(TRAIN \times CDV)$. The sub-components invariants are derived from the different initial abstract models (see their labels in Figure 7.18). Invariants that only restrain the sub-component variables are automatically included although additional ones can be added manually.

7.4.7. *Machine Train*

Machine *Train* models the trains in the metro system. Trains entering/leaving a section, modeled by events $enterCDV$ and $leaveCDV$, are part of this sub-component, (see Figure 7.19b). The interaction with sub-component *Track* occurs through parameters $t1$, $c1$ and $c2$ (see events $Track.leaveCDV$ in Figure 7.18). Variables *door* and $door\_state$ are part of this sub-component and consequently, the events that modify these variables: $openDoor$ and $closeDoor$. Moreover, since the emergency button is part of a train, the respective variable $emergencyButton$ (and the modification event $toggleEmergencyButton$) is also included in this sub-component. Event $recvTrain$ Msg receives messages sent to the trains and the content is stored in the variable *permit*. Although variable $permit$ is set based on the content of the messages exchanged between *Train* and *Track*, that variable is read by trains. This is the reason why it is allocated to this sub-component. The events that change the speed of the train are also included in this sub-component: $brake$, $stopBraking$, $changeSpeed$ due to variables $speed$ and $braking$ as depicted in Figure 7.19.

7.4.8. *Machine Middleware*

Finally, the communication layer is modeled by *Middleware* as seen in Figure 7.20. *Middleware* bridges *Track* and *Trains*, by receiving messages ($sendTrainMsg$) from the tracks and delivering to the trains ($recvTrainMsg$). Variable $tmsgs$ is used as a buffer.

Benefiting from the monotonicity of the shared event approach, the resulting sub-components can be further refined.

7.4.9. *Refinement of Train: Train_M1*

In Train_M1, carriages are introduced as parts of a train. Each carriage has an individual alarm, which when activated, triggers the train alarm (enables the emergency button of the train). Each train has a limited number of carriages. Each carriage has a set of doors and the sum of carriage doors corresponds to the doors of a train. The properties to be preserved are as follows:

```
machine Train sees MetroSystem_C1

variables trns speed permit braking emergency_button door_state door

invariants
  theorem @typing_trns trns ∈ ℙ(TRAIN)
  theorem @typing_door_state door_state ∈ ℙ(DOOR × DOOR_STATE)
  theorem @typing_braking braking ∈ ℙ(TRAIN)
  theorem @typing_speed speed ∈ ℙ(TRAIN × ℤ)
  theorem @typing_permit permit ∈ ℙ(TRAIN × BOOL)
  theorem @typing_door door ∈ ℙ(TRAIN × ℙ(DOOR))
  theorem @typing_emergency_button emergency_button ∈ ℙ(TRAIN × BOOL)
  @MetroSystem_M0_inv3 trns ⊆ TRAIN
  @MetroSystem_M0_inv9 braking ⊆ trns
  @MetroSystem_M0_inv10 speed ∈ trns → ℕ
  @MetroSystem_M1_inv2 permit ∈ trns → BOOL
  @MetroSystem_M1_inv7 emergency_button ∈ trns → BOOL
  @MetroSystem_M2_inv1 door_state ∈ DOOR → DOOR_STATE
  @MetroSystem_M2_inv2 door ∈ trns → ℙ(DOOR)
  @MetroSystem_M2_inv3 ∀t1,t2·t1 ∈ dom(door) ∧ t2 ∈ dom(door) ∧ t1 ≠t2 ⇒ door(t1) ∩ door(t2) = ∅
  @MetroSystem_M2_inv4 ∀t·t ∈ dom(door) ⇒(∃d·d⊆door(t) ∧ door_state[d]={OPEN} ⇒ speed(t)=0)
  @MetroSystem_M2_inv5 ∀t·t ∈ dom(door) ∧ speed(t) > 0 ⇒ door(t) ⊆ door_state∼[{CLOSED}]
  theorem @MetroSystem_M2_thm1 ∀t·t ∈ dom(door) ∧ door(t) ∩ door_state∼[{OPEN}] =∅
                               ⇒ door(t)⊆door_state∼[{CLOSED}]
```

(a) Variables and invariants in *Train*

```
event recvTrainMsg
  any t1 bb
  where
    @typing_t1 t1 ∈ TRAIN
    @typing_bb bb ∈ BOOL
  then
    @act2 permit(t1)≔bb
end

event changeSpeed
  any t1 s1
  where
    @typing_t1 t1 ∈ TRAIN
    @typing_s1 s1 ∈ ℤ
    @grd1 s1 ∈ ℕ
    @grd2 t1 ∈ dom(door)
    @grd3 t1 ∈ braking ⇒ s1 < speed (t1)
    @grd4 door(t1) ∩ door_state∼[{OPEN}] =∅
  then
    @act1 speed (t1) ≔ s1
end

event brake
  any t1
  where
    @typing_t1 t1 ∈ TRAIN
    @grd1 t1 ∈ trns\braking
    @grd2 t1 ∈ dom(emergency_button)
    @grd3 permit(t1) = FALSE
          ∨ emergency_button(t1)=TRUE
  then
    @act1 braking ≔ braking ∪ {t1}
end

event openDoor
  any t occpTrns ds
  where
    @typing_t t ∈ TRAIN
    @typing_occpTrns occpTrns ∈ ℙ(CDV)
    @typing_ds ds ∈ ℙ(DOOR)
    @grd1 t ∈ dom(door)
    @grd2 speed(t) = 0
    @grd4 occpTrns ∩ PLATFORM ≠ ∅
          ∨ emergency_button(t) = TRUE
    @grd5 ds ⊆ door(t)
    @grd6 ∃d·d∈ds⇒door_state(d)=CLOSED
    @grd7 ds≠∅
  then
    @act1 door_state≔ door_state ◁ (ds×{OPEN})
end

event closeDoor
  any t ds
  where
    @typing_t t ∈ TRAIN
    @typing_ds ds ∈ ℙ(DOOR)
    @grd1 t ∈ dom(door)
    @grd2 speed(t) = 0
    @grd3 ds ⊆ door(t)
    @grd4 door_state[ds]={OPEN}
    @grd5 ds≠∅
  then
    @act1 door_state≔ door_state ◁ (ds×{CLOSED})
end

event toggleEmergencyButton
  any t value occpTrns
  where
    @typing_t t ∈ TRAIN
    @typing_occpTrns occpTrns ∈ ℙ(CDV)
    @grd1 t ∈ dom(door)
    @grd2 value ∈ BOOL
    @grd3 door(t) ∩ door_state∼[{OPEN}] ≠ ∅
          ∧ PLATFORM ∩ occpTrns=∅
          ⇒ value = TRUE
  then
    @act1 emergency_button(t)≔ value
end

event addDoorTrain
  any t d
  where
    @typing_d d ∈ ℙ(DOOR)
    @typing_t t ∈ TRAIN
    @grd1 t ∈ trns
    @grd2 d ⊆ DOOR
    @grd3 ∀tr·tr∈dom(door) ∧ tr≠t
          ∧ door(tr)≠∅ ⇒ d∩door(tr)=∅
    @grd5 speed(t)=0
    @grd7 d∩door(t)=∅
  then
    @act1 door(t)≔door(t)∪d
    @act2 door_state≔door_state◁(d×{CLOSED})
end

event removeDoorTrain
  any t d
  where
    @typing_d d ∈ ℙ(DOOR)
    @typing_t t ∈ TRAIN
    @grd1 t ∈ dom(door)
    @grd2 d ⊆ DOOR
    @grd3 d ⊆ door(t)
    @grd4 door_state[d]={CLOSED}
    @grd5 speed(t)=0
  then
    @act1 door(t) ≔ door(t)\d
end

event leaveCDV
  any t1 c1 c2
  where
    @typing_t1 t1 ∈ TRAIN
    @grd1 t1 ∈ dom(door)
    @grd2 c1 ∈ CDV
    @grd3 c2 ∈ CDV
    @grd4 speed(t1)>0
    @grd11 door(t1)∩door_state∼[{OPEN}]=∅
    @grd12 permit(t1)=TRUE
end
```

(b) Some events of *Train*

Figure 7.19. *Excerpt of Train*

1. There is a limit to the number ($MAX\_NUMBER\_CARRIAGE$) of carriages per train.

2. Whenever a carriage alarm is activated, then the emergency button of that same train is activated.

3. The sum of carriage doors corresponds to the doors of a train.

```
machine Middleware sees MetroSystem_C1

variables tmsgs

invariants
  theorem @typing_tmsgs tmsgs ∈ ℙ(TRAIN × ℙ(BOOL))

events
  event INITIALISATION
    then
      @act1 tmsgs ≔ ∅
  end

  event sendTrainMsg
    any t1 bb
    where
      @typing_t1 t1 ∈ TRAIN
      @typing_bb bb ∈ BOOL
      @grd1 t1 ∈ dom(tmsgs)
      @grd2 tmsgs(t1)=∅
    then
      @act1 tmsgs(t1) ≔ {bb}
  end

  event recvTrainMsg
    any t1 bb
    where
      @typing_t1 t1 ∈ TRAIN
      @typing_bb bb ∈ BOOL
      @grd1 t1 ∈ dom(tmsgs)
      @grd2 bb ∈ tmsgs(t1)
    then
      @act1 tmsgs(t1)≔∅
  end

  event addTrain
    any t oc
    where
      @typing_t t ∈ TRAIN
      @grd1 oc ∈ CDV
    then
      @act6 tmsgs(t)≔∅
  end
```

Figure 7.20. *Machine Middleware*

The definition of these requirements need the introduction of some static elements, such as a carrier set *CARRIAGE*, constants *MAX_NUMBER_CARRIAGE*, and *DOOR_CARRIAGE* (function between *DOOR* and *CARRIAGE*). The latter is defined as a constant because the number of doors in a carriage does not change. Context *Train_C2* is depicted in Figure 7.21a. Several variables are added, such as $train\_carriage$ relating carriages with trains and $carriage\_alarm$ that is a total function between $CARRIAGE$ and BOOL, illustrated in Figure 7.21b. Property 1 is expressed by invariant *inv6* stating that trains have a maximum of *MAX_NUMBER_CARRIAGE* carriages. Property 2 is defined in *inv7* as seen in Figure 7.21b. Events *activateEmergencyCarriageButton* and *deactivateEmergency TrainButton* refine abstract event *toggleEmergencyButton*: the first event enables a carriage alarm and consequently enables the emergency button of the train; the latter occurs when the emergency button of a train is active and corresponds to the deactivation of the last enabled carriage alarm, which results in deactivating the emergency button; a new event $deactivateEmergencyCarriageButton$ is added to model the deactivation of a carriage alarm when there is still another alarm enabled for the same train (guards $grd4$ and $grd5$). The allocation and removal of carriages (events *allocateCarriageTrain*

```
context Train_C1 extends MetroSystem_C1

constants MAX_NUMBER_CARRIAGE
          DOOR_CARRIAGE

sets CARRIAGE

axioms
  @axm1 MAX_NUMBER_CARRIAGE ∈ ℕ1
  @axm2 DOOR_CARRIAGE ∈ DOOR→CARRIAGE
  @axm3 ∀c·c∈ran(DOOR_CARRIAGE)
                  ⇒DOOR_CARRIAGE∼[{c}]≠∅
end
```

(a) Context *Train_C1*

```
machine Train_M1 refines Train  sees Train_C1

variables trns speed permit braking door_state door emergency_button
        train_carriage carriage_alarm

invariants
  @inv1 finite(trns)
  @inv2 carriage_alarm ∈ CARRIAGE → BOOL
  @inv3 train_carriage ∈ CARRIAGE ↠ trns
  @inv4 finite(train_carriage)
  @inv5 finite(dom(train_carriage))
  @inv6 ∀t·t ∈ trns ⇒ card(train_carriage∼[{t}])≤MAX_NUMBER_CARRIAGE
  @inv7 ∃c·(c ∈ dom(train_carriage) ∧ carriage_alarm(c) = TRUE
  ⇔ c ∈ dom(train_carriage) ∧ emergency_button(train_carriage(c))= TRUE)
  @inv8 ∀t·t∈dom(door) ⇒ door(t)=DOOR_CARRIAGE∼[train_carriage∼[{t}]]
```

(b) Variables and Invariants of *Train_M1*

```
event activateEmergencyCarriageButton
refines toggleEmergencyButton
  any c occpTrns
  where
    @grd1 occpTrns ∈ ℙ(CDV)
    @grd2 c ∈ dom(train_carriage)
    @grd3 carriage_alarm(c) = FALSE
  with
    @value value = TRUE
    @t t = train_carriage(c)
  then
    @act1 carriage_alarm(c) ≔ TRUE
    @act2 emergency_button(train_carriage(c)) ≔ TRUE
end

event deactivateEmergencyCarriageButton
  any c
  where
    @grd1 c ∈ dom(train_carriage)
    @grd2 emergency_button(train_carriage(c)) = TRUE
    @grd3 carriage_alarm(c) = TRUE
    @grd4 {c} ≠ (dom(carriage_alarm ⊳ {TRUE})
              ∩ train_carriage∼[{train_carriage(c)}])
    @grd5 card(train_carriage⊳{train_carriage(c)})>1
  then
    @act1 carriage_alarm(c)≔ FALSE
end

event deactivateEmergencyTrainButton
refines toggleEmergencyButton
  any c occpTrns
  where
    @grd1 occpTrns ∈ ℙ(CDV)
    @grd2 c ∈ dom(train_carriage)
    @grd3 emergency_button(train_carriage(c)) = TRUE
    @grd4 carriage_alarm(c) = TRUE
    @grd5 {c} = (dom(carriage_alarm ⊳ {TRUE})
               ∩ train_carriage∼[{train_carriage(c)}])
    @grd6 door(train_carriage(c))∩door_state∼[{OPEN}] = ∅
  with
    @value value = FALSE
    @t t = train_carriage(c)
  then
    @act1 carriage_alarm(c)≔ FALSE
    @act2 emergency_button(train_carriage(c)) ≔ FALSE
end

event alocateCarriageTrain refines addDoorTrain
  any c t
  where
    @grd1 c ∈ CARRIAGE\dom(train_carriage)
    @grd2 carriage_alarm[{c}]= {FALSE}
    @grd3 ∀tr·tr∈dom(door) ∧ tr≠t ∧ door(tr)≠∅
          ⇒ DOOR_CARRIAGE∼[{c}]∩door(tr)=∅
    @grd4 t ∈ trns
    @grd5 emergency_button(t) = FALSE
    @grd6 finite(train_carriage∼[{t}])
    @grd7 card(dom(train_carriage ⊳ {t}))
                              <MAX_NUMBER_CARRIAGE
    @grd8 speed(t)=0
    @grd9 DOOR_CARRIAGE∼[{c}] ∩ door(t)=∅
  with
    @d d=(DOOR_CARRIAGE∼[{c}])
  then
    @act1 train_carriage(c)≔ t
    @act2 door(t)≔door(t) ∪ DOOR_CARRIAGE∼[{c}]
    @act3 door_state≔
      door_state<+(DOOR_CARRIAGE∼[{c}]×{CLOSED})
end

event removeCarriageTrain refines removeDoorTrain
  any c t
  where
    @grd1 t ∈ dom(door)
    @grd2 c↦t ∈ train_carriage
    @grd3 carriage_alarm(c) = FALSE
    @grd4 emergency_button(t) = FALSE
    @grd5 speed(t)=0
    @grd6 DOOR_CARRIAGE∼[{c}]⊆door(t)
    @grd7 DOOR_CARRIAGE∼[{c}]≠∅
    @grd8 door_state[DOOR_CARRIAGE∼[{c}]]={CLOSED}
  with
    @d d = (DOOR_CARRIAGE∼[{c}])
  then
    @act1 train_carriage ≔ {c}⩤train_carriage
    @act2 door(t)≔door(t)\DOOR_CARRIAGE∼[{c}]
end
```

(c) Some events of *Train_M1*

Figure 7.21. *Excerpt of machine Train_M1*

and *removeCarriageTrain*) refine *addDoorTrain* and *removeDoorTrain*, respectively. In these two events, the parameter d representing a set of doors is replaced in the witness section by the doors of the added/removed carriage: $d = DOOR\_CARRIAGE^{-1}[\{c\}]$.

7.4.10. *Further development*

Details of some remaining refinement and decomposition steps may be found in [SIL 12]. *Carriage* is refined and decomposed until it fits in a generic model *GCDoor* corresponding to a *Generic Carriage Door* development as seen in Figure 7.22. A generic model *GCDoor* is instantiated into two instances: *EmergencyDoors* and *ServiceDoors* benefiting from the refinements in the pattern.

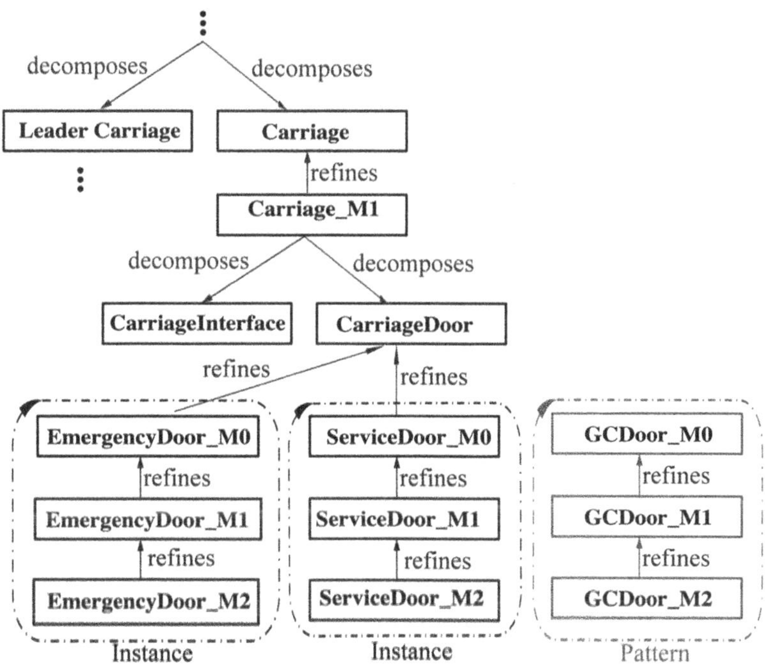

Figure 7.22. *Carriage refinement diagram and door instantiation*

7.4.11. *Conclusion*

We modeled a metro system case study, starting by proving its global properties through several refinement steps. Afterwards due to an architectural decision and to alleviate the problem of modeling and handling a large system in one single machine, the system is decomposed in three sub-components. We further refine one of the resulting sub-components (*Train*), introducing several details in four refinement levels.

The derivation of the distributed rail system illustrates a formal design approach for embedded controllers that takes into account models of the physical behavior as well as required control behavior. Traditionally, formal methods are used to verify correctness of computer systems with respect to a specification. Here, we are using formal methods to model and reason about a system as a whole, both the physical system and the required control behavior.

Specifying the system-level model does require skill in deciding on the appropriate abstractions, what aspects of behavior need to be modeled and what aspects can be left out of the model. The benefit of the system-level model is that it is easier to understand and reason about the behavior of the system as a whole.

The complexity of the entities and the relationships between them is handled through the use of refinement, which allows complexity to be introduced and reasoned about in steps. We made use of refinement for two main purposes, to introduce communication mechanisms leading to system partition and to replace abstract structures by more concrete realizations (such as replacing $next$ by pos).

Although we are mainly interested in safety properties, the model checker ProB [WIK 02] proved to be very useful as a complementary tool during the development of this case study. In some stages of the development, all the proof obligations were discharged but with ProB, we discovered that the system was deadlocked due to some missing detail. In large developments, these situations possibly occur more frequently. Therefore, we suggest discharging the proof obligations to ensure the safety properties are preserved and running the ProB model checker to confirm that the system is free from deadlocks.

7.5. Acknowledgments

The authors of this chapter would like to acknowledge funding from the FP7 DEPLOY Project (www.deploy-project.eu).

7.6. Bibliography

[ABR 96] ABRIAL J.-R., *The B-book: Assigning Programs to Meanings*, Cambridge University Press, New York, USA, 1996.

[ABR 07] ABRIAL J.-R., HALLERSTEDE S., "Refinement, decomposition, and instantiation of discrete models: application to Event-B", *Fundamenta Informaticae*, vol. 77, no. 1-2, p. 1-28, 2007.

[ABR 08] ABRIAL J.-R., *Modeling in Event-B: System and Software Engineering*, Cambridge University Press, 2008.

[ABR 10] ABRIAL J.-R., BUTLER M., HALLERSTEDE S., HOANG T.S., MEHTA F., VOISIN L., "Rodin: an open toolset for modelling and reasoning in Event-B", *STTT*, vol. 12, no. 6, p. 447-466, 2010.

[BAC 89] BACK R.-J., "Refinement calculus, part II: parallel and reactive programs", *REX Workshop*, p. 67-93, 1989.

[BUT 97] BUTLER M.J., "An approach to the design of distributed systems with B AMN", *ZUM*, p. 223-241, 1997.

[BUT 02] BUTLER M., "A system-based approach to the formal development of embedded controllers for a railway", *Design Automation for Embedded Systems*, vol. 6, p. 355-366, 2002.

[BUT 09] BUTLER M., "Decomposition structures for Event-B", *Integrated Formal Methods iFM2009, Springer LNCS 5423*, February 2009.

[DAM 08] DAMCHOOM K., BUTLER M., ABRIAL J.-R., "Modelling and proof of a tree-structured file system in Event-B and Rodin", *Proceedings of the 10th International Conference on Formal Methods and Software Engineering, ICFEM '08*, p. 25-44, Springer-Verlag, Berlin, Heidelberg, 2008.

[EVE] Event-B and Rodin Website, http://www.event-b.org/.

[FAL 09] FALAMPIN J., BUTLER M., FITZGERALD J., Deploy deliverable d16 d2.1 pilot deployment in transportation (wp2). http://www.deploy-project.eu/pdf/D16_final6, September 2009.

[LEU 08] LEUSCHEL M., BUTLER M., "ProB: an automated analysis toolset for the B Method", *International Journal on Software Tools for Technology Transfer (STTT)*, vol. 10, p. 185-203, 2008.

[SIL 11] SILVA R., PASCAL C., HOANG T.S., BUTLER M., "Decomposition tool for Event-B", *Software, Practice and Experience*, vol. 41, no. 2, p. 199-208, 2011.

[SIL 12] SILVA R., Supporting Development of Event-B Models, Draft PhD Thesis, University of Southampton, 2012.

[SNO 06] SNOOK C.F., BUTLER M.J., "UML-B: formal modeling and design aided by UML", *ACM Transactions on Software Engineering and Methodology*, vol. 15, no. 1, p. 92-122, 2006.

[TSD 12] Transportation Systems Design Inc, Communications based train control, http://www.tsd.org/cbtc/, January 2012.

[WIK 01] Wiki. B2Latex [Online]. http://wiki.event-b.org/index.php/B2Latex.

[WIK 02] Wiki. ProB [Online]. http://wiki.event-b.org/index.php/ProB.

[WIK 03] Wiki. UML-B [Online]. http://wiki.event-b.org/index.php/UML-B.

Chapter 8

Conclusion

8.1. Introduction

Formal techniques (simulation, *model-checking*, abstract interpretation, proof, etc.) are not new. Indeed, the first papers that cover the subject date from the 1970s (for examples, see [HOA 69, COU 77, DIJ 76]). However, the implementation of formal methods dates back to the 1980s [SPI 89, JON 90, HAL 91], with industrial use starting in the 1990s [BEH 93, ARA 97].

In [BOW 95, ARA 97], we find the early feedback from industrialists concerning formal techniques and, in particular, feedback on the B method [ABR 96, BEH 93, BEH 96, BOU 06], the language LUSTRE [HAL 91, ARA 97], SAO+ predecessor[1] of SCADE [BEN 03, DOR 08]. Other works, such as [MON 00, HAD 06], provide a panorama of the formal methods in a more scientific approach.

Within the framework of this book, we presented various industrial uses of formal methods. These uses cover various fields (railway, automotive, aeronautics, nuclear power, etc.).

In [BOU 12a] *Static Analysis of Software*, we presented some formal tools based on the static analysis and abstract interpretation of code. In [BOU 12b], we presented some approaches based on the modelization (SCADE, B Method,

---

Chapter written by Jean-Louis BOULANGER.

1 It should be noted that initially SCADE was a development environment based on the language LUSTRE and that since Version 6, SCADE became a language of its own (the code generator for Version 6 works well inputting a SCADE model but not a LUSTRE code).

Mathlab/Simulink, and ControlBuild). In this book, several environments and formal methods have been presented, including SPARK Ada, AltaRica, Polyspace®, Escher Verification Studio Perfect Developer, Event-B and Rodin.

The goal of this chapter is to summarize what was presented and to provide an update on the changes taking place. Recall that the objective of these three books[2] is to make an inventory of the use of *formal techniques* (simulation, *model-checking* [BAI 08], proof, abstract interpretation, etc.) and *formal methods* (for example, B method [ABR 96], LUSTRE [HAL 91, BEN 03, HAL 05], SCADE [HAL 05, DOR 08], AltaRica, SPARK Ada [BAR 03], etc.) by industrialists. We will therefore not discuss the scientific challenges and possible future developments.

8.2. Structured, semi-formal and/or formal methods

8.2.1. *E/E/PE system*

The CEI/IEC 61508 [IEC 98] standard is a generic standard, which may be applied to all electronically based and/or programmable electronic complex systems termed E/E/PE (electric/electronic/programmable electronic). This standard is edited by the *International Electrotechnical Commission*[3] (IEC), which is the international standardization organization responsible for electricity, electronics and related techniques.

The CEI/IEC 61508 [IEC 98] standard proposes a globalized approach to security, in the sense of safety/verifiable safety[4], comparable to the ISO 9001:2008 [ISO 08] system for quality. The CEI/IEC 61508 [IEC 98] standard is totally consistent with the convergence observable between different industrial sectors (aeronautical, nuclear, rail, automobile, manufacturer, etc.), but the content of the CEI/IEC 61508 standard is sufficiently complex, not to mention unusual, as to need "to be guided through", as interested readers will be able to read in [ISA 05, SMI 07].

---

2 The French version of the *Industrial implementation of formal methods* series is composed of four volumes, whereof [BOU 11a, BOU 11b, BOU 12c, BOU 12d]; and three books in English (this book, along with [BOU 12a, BOU 12b].

3 To find out more about the IEC, visit: www.iec.ch.

4 The CEI/IEC 61508 [IEC 98] standard does not cover the confidentiality and/or integrity aspects, which are linked to establishment of precautions aiming to prevent non-authorized persons from damaging and/or affecting safety achieved by the E/E/PE system or systems. Notably, management of the network aspect so as to avoid intrusions comes to mind here.

The CEI/IEC 61508 [IEC 98] standard defines the notion of SIL (*Safety Integrity Level*). The SIL allows the security level of a system to be quantified. Safety integrity level 1 (SIL1) is the lowest and safety integrity level 4 (SIL4) is the highest.

The CEI/IEC 61508 standard details the requirements necessary to be implemented to create a system and/or software, which achieves its safety objectives (SIL). The higher the safety level to be obtained, the more severe the requirements to be implemented and the greater the financial impact. In effect, to obtain a low probability for a dangerous failure, greater financial investment is necessary.

The SIL (*Safety Integrity Level*) enables prescriptions concerning safety functions integrity to allocate to E/E/PE systems to be specified.

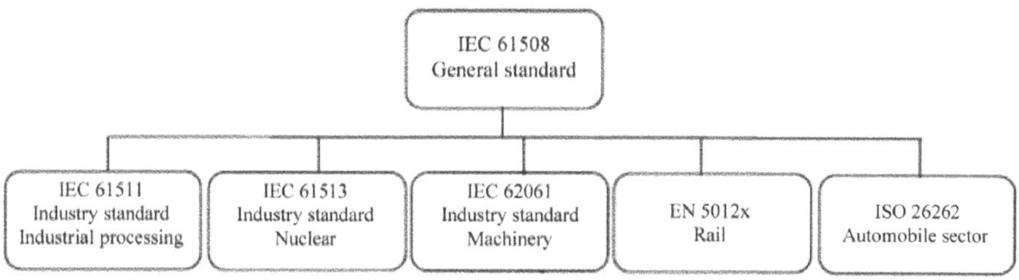

Figure 8.1. *The CEI/IEC 61508 [IEC 98] standard and these declensions*[5]

For several years, the generic CEI/IEC 61508 [IEC 98] and daughter standards cover different sectors, as Figure 8.1 shows.

The CEI/IEC 61508 [IEC 98] standard is greatly used for the development of "*safety-critical*" software in the automation and automobile sectors, as well as for installation of industrial process controls. Part 3 of the CEI/IEC 61508 standard is dedicated to production of software applications.

8.2.2. *Rail sector*

Today, rail projects are governed by some texts (decrees, orders, etc.) and a normative standard (CENELEC[6] EN 50126 [CEN 00], EN 50129 [CEN 03] and

---

5 Figure 8.1 reveals a link between the CEI/IEC 61508 standard and the CEI/IEC 61513 [IEC 01] standard, but this link is incorrect. In effect, nuclear standards existed well before the CEI/IEC 61508 standard and the link is only a convention.

6 CENELEC for *Comité européen de normalisation électrotechnique* (European Elecrotechnical Standardization Committee), see: www.cenelec.eu.

EN 50128 [CEN 01a]) aiming to define and achieve certain FMDS (fiabilité, maintenabilité, disponibilité et sécurité = *reliability, maintainability, availability, and safety*) objectives.

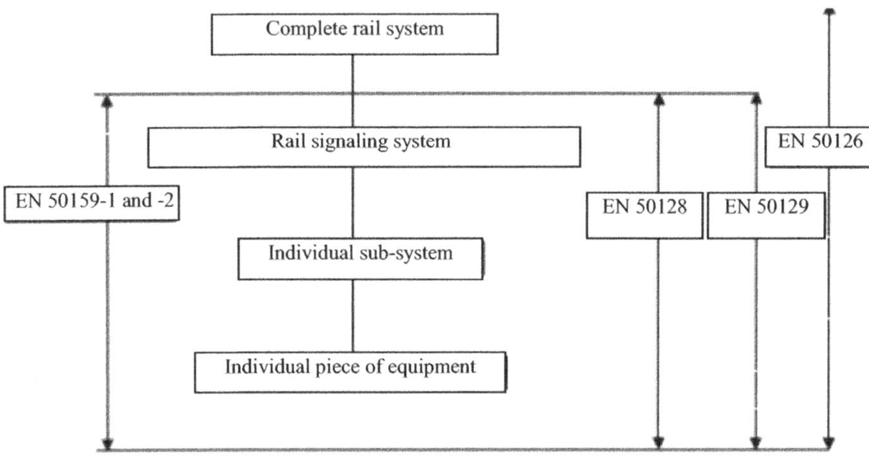

Figure 8.2. *CENELEC standard*

Three standards (see Figure 8.2) make it possible to cover aspects concerning safety/verifiable *safety* of the system, right down to the hardware and/or software elements.

For the notion of safety-confidentiality, the CENELEC standard is completed by the EN 50159-1 [CEN 01b] and EN 50159-2 [CEN 01c] standards, which are applicable if the application uses respectively some closed and open network.

These standards now have a European statute, applying to all the member states of the EU, and have been endorsed by the *International Electrotechnical Commission*; application of CEI/IEC 61508 [IEC 98] to the rail sector also conferring an international statute on them.

Safety, as an element of *safe functioning*, is obtained by establishment of designs, methods, tools, all through the lifecycle. A safety study necessitates analysis of the failures of the system. It must set out to identify and quantify severity of the potential consequences and as far as possible, the predictable frequency of these failures.

Among the risk reduction actions enabling achievement of an acceptable level of risk, the CENELEC EN 50129 [CEN 03] standard describes allocation of safety objectives to the functions of the system and its sub-systems. The safety integrity level (SIL) is defined as one of the discrete levels for specifying safety integrity requirements of safety functions allocated to safety systems. The SIL goes from the value of 1 up to the value of 4.

The notion of SIL in the sense of the standard CENELEC EN 50129 [CEN 03] is not the same as in the standard CEI/IEC 61508 [IEC 98]. The differences concern the identification method and the impact. Usage, which is made within the standards, is, however, identical.

The CENELEC EN 50128 [CEN 01a] standard is specifically dedicated to the development aspects of software for the rail sector. Regarding software, the SSIL (Software SIL) makes it possible to define different levels of criticality: from 0 (no danger) to 4 (critical). The CENELEC EN 50128 [CEN 01a] standard specifies *the procedures and the technical prescriptions* applicable for the development of programmable electronic systems used in applications for rail command and protection. The CENELEC EN 50128 standard is thus not normally applicable to all software applications for the rail sector. The EN 50155 [AFN 01] standard is normally applicable to all the embedded applications in a train. However, the EN 50155 standard calls for the CENELEC EN 50128 standard.

The various chapters in this book gave us the opportunity to present examples in the industrial environment using formal methods. Even if, to date, there are several environments and a significant number of applications, we should note that there was a change. We passed from a *timid* implementation of formal techniques to an *industrial* implementation as will be explained later.

However, this change has not occurred without posing some difficulties. The development of a software application impacting safety is a rather difficult endeavor that is oriented towards *quality control*[7].

Indeed, the only effective technique for creating a safety software is to avoid introducing defects and/or to detect the maximum amount of defects. To do this, quality must be controlled. Quality control by the definition of process and the considerable production of document.

Introducing formal methods into the process makes us pass from a *quality control* approach to a *model-centered* approach that can give the impression that

7 Quality control, in the sense of applying a process that conforms to a standard concerning quality control, for example the ISO 9001:2008 [ISO 08].

documentation becomes useless and this produces a software in accordance with the safety integrity level that only depends of the safety integrity level of the tools used.

8.2.3. *Taking into account techniques and formal methods*

It is interesting to note that the definitions of the B and C annexes (part 7 of the CEI/IEC 61508 standard) use some articles and/or books, which are dated between 1970 and 1998, as references. It seems necessary to bring this annex up to date.

8.2.3.1. *Definitions*

8.2.3.1.1. Semi-formal method

According to the CEI/IEC 61508 (B.2.3) standard, a *semi-formal method* offers a means of developing a description of a system at a stage of development (specification, architecture and/or design). The description can be analyzed or animated so as to verify that the system modeled satisfies the requirements (real and/or specified). It is indicated that state diagrams (finite controllers) and temporal Petri nets are two semi-formal methods.

The CENELEC EN 50128 standard identifies semi-formal methods (see Table 8.1). In the outline of the new CENELEC EN 50128 [CEN 11] standard, semi-formal methods as techniques have disappeared. This is due to the ambiguity that existed between formal and semi-formal and between semi-formal and structured methods. Certain manufacturers consider the existence of a modeling tool as enough to demand establishment of a semi-formal method.

	SSIL0	**SSIL1**	**SSIL2**	**SSIL3**	**SSIL4**
Logic/function block diagram	R	R	R	HR	HR
Sequence diagrams	R	R	R	HR	HR
Data flow diagrams	R	R	R	R	R
Finite state controllers/state transition schematic	-	R	R	HR	HR
Temporal petri nets	-	R	R	HR	HR
Decision/truth table	R	R	R	HR	HR

Table 8.1. *CENELEC EN 50128 [CEN 01] – Table A.18*

Whereas the existence of a tool has only a slight link with the existence of a semantic; see for example, the Labview environment, which, although graphic, does not use a semantic. The question is all the more relevant for environments based on UML notation.

8.2.3.1.2. Formal method

Section B.30 of the CEI/IEC 61508 standard indicates that a formal method (HOL, CSP, LOTOS, CCS, linear temporal logic (LTL), VDM[8] [JON 90], and Z[9] [SPI 89]) enables an unambiguous and coherent description of a system at a stage of development (specification, architecture, and/or design) to be produced. The description takes a mathematical form and can be subjected to a mathematical analysis. This mathematical analysis may be performed automatically.

A formal method generally offers a notation, a technique for processing a description in this notation, and a verification process for controlling correction of the requirements.

The CEI/IEC 61508 standard indicates that it is possible to make transformations right down to "a logic circuit design"[10].

Petri nets and state machines (mentioned in the outline of semi-formal methods) can be considered as formal methods, according to the degree of conformity to a rigorous mathematical basis of their uses.

8.2.3.1.3. Structured method

The aim of *structured methods* (EN 50128, B.60) is to promote quality of software development in bringing attention to the first phases of the lifecycle.

This type of method calls for some precise and intuitive notations (generally computer-aided) so as to produce and document the requirements and characteristics of installation of the product.

Structured methods are tools for reflection. They are based on an order of logical reflection, an analysis of the system taking into account environment, production, and documentation of the system, breakdown of data and of the functions and the

8 In the IEC 61508 standard, the references to VDM include VDM++ [DUR 92], which is a real-time and object-oriented extension of VDM. To find out more, visit: www.vdmportal.org.
9 In the IEC 61508 standard, the B method [ABR 96] is seen as a method associated with Z.
10 These are the terms of the CEI/IEC 61508 standard.

elements to verify; they must impose a weak intellectual service (simple, intuitive, and pragmatic method).

Among the structured methods are SADT [LIS 90], JSD, CORE, Yourdon real-time, and MASCOT.

Structured specification is a technique, which aims to give prominence to the simplest visible relationships between the partial requirements and the functional specification. Analysis is refined by stages, right down to obtaining small and straightforward partial requirements. A hierarchical structure of partial requirements is then obtained, which will allow the total requirements to be specified. The method highlights interfaces between requirements and enables interface failures to be avoided.

8.2.3.2. *Computer-aided specification tools*

In section B.24 of the outline of the CEI/IEC 61508 standard, it is indicated that implementation of computer-aided specification tools allows a specification to be obtained in the form of a database, which can be automatically inspected to evaluate coherence and completion. The tool can enable animation.

The application can be made at any of the different phases: specification, architecture, and/or design. For the design phase, use is recommended once there are tools, but it will be necessary to show correction of the tool (experience feedback and/or independent verification of the results).

8.2.4. *Software requirement specification*

8.2.4.1. *Requirement management*

The standards applicable to different industries render identification of the software application requirements mandatory. It should be noted that the literature presents several terms: prescriptions, recommendation, requirement, constraint, and property, which seem equivalent. The general CEI/IEC 61508 [IEC 98] standard uses the notion of prescription, but the most-used term remains the notion of requirement.

Standards recommend that the specification of a software application be composed from a textual description of the requirements and all the notations necessary to facilitate understanding of the requirement. So, for the EN 50128 [CEN 01] standard, there is Table A.2 (Table 8.2).

DEFINITION 8.1.– PROPERTY. A property is a characteristic that a (hardware or software) component must verify.

The properties are conventionally separated into two groups: properties of safety (see definition 8.2) and properties of liveness (see definition 8.3).

	SSIL0	SSIL1	SSIL2	SSIL3	SSIL4
Formal methods comprising for example CCS, CSP, HOL, LOTOS, OBJ, VDM, Z, B	-	R	R	HR	HR
Semi-formal methods	R	R	R	HR	HR
Structured methodology comprising for example JSD, MASCOT, SADT, SDL, SSADM and YOURDON	R	HR	HR	HR	HR

Table 8.2. *CENELEC EN 50128 [CEN 01] – Table A.2*

DEFINITION 8.2.– PROPERTY OF SAFETY. The property of safety asserts that something bad never occurs during execution.

As an example of the property of safety, we can cite the absence of blocking, mutual exclusion, etc. and in our case, the absence of a collision between trains.

DEFINITION 8.3.– PROPERTY OF LIVENESS. The property of *liveness* expresses that something good inevitably occurs during execution.

The termination of an application, the absence of famine (constant progression of applications), and the guarantee of service are many examples of properties of liveness. It is possible to show that any property P can be expressed as the conjunction of the property of liveness V and of the property of safety S.

As we presented it, the concept of property is related to an identification *a posteriori* of need that the application being analyzed must respect. In this context, the properties are identified in the phase of "conception" by analyzing the element in entries, which are called requirement. In the specification phase, all the software requirements are identified and labeled.

The difficulty resides then in the definition of the concept of a requirement. There are several studies that attempt to identify what a requirement is and how to account for it. [HUL 05] presents one of the most complete syntheses. We revisit the following definition, which is the result of work led by industrialists.

DEFINITION 8.4.– REQUIREMENT[11]. A requirement is a statement that translates a need and/or constraints (techniques, costs, time, etc.). This statement is written in a language that can be natural, mathematical, etc.

To get a clear identification, a requirement is a labelized element (the label allows a single identification) that characterizes an element of the system to be carried out. For each requirement, it is necessary to know the source.

Within the framework of the execution, the requirements (definition 8.5) will be analyzed and taken into account during each stage of the cycle of the system execution. In the context of the phases of hardware and/or software design, a requirement could be translated in the form of properties (definition 8.2) carried on the components (software or hardware).

Remember (Figure 8.3) that there are two families of requirements, the functional ones that are related to the behavior of the system, and the non-functional ones (also called extra-functional requirements) that are related to quantifiable objectives (performance, level of safety, reliability, availability, etc.). In Chapter 2 of [RAM 09] and Chapter 3 of [RAM 11], we have presented some examples of requirements management in the automobile and rail industries.

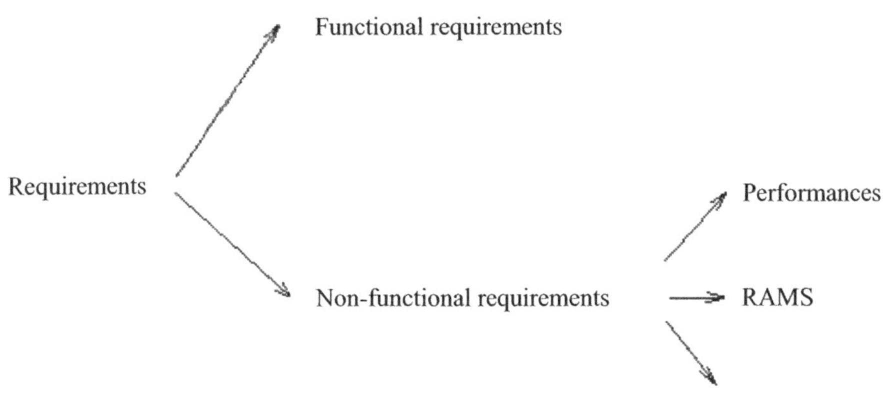

Figure 8.3. *Different types of requirements*

Table 8.3 is an extract from [STAN 94], which allows for emphasis of the fact that over 30% of the causes of failure in production originate from incomplete requirements, incomplete needs descriptions, and unrealistic needs.

11 AFIS is the French Association of System Engineering. One of its areas of work is specifically dedicated to managing requirements. For more information: www.a fi s.fr.

Description	%
Incomplete Requirement	13.1%
Lack of user involvement	12.4%
Lack of Resources	10.6%
Unrealistic expectations	9.9%
Lack of executive support	9.3%
Changing Requirement/specification	8.7%
Lack of planning	8.1%
Did not need it any longer	7.5%

Table 8.3. *Distribution of the causes of failure*

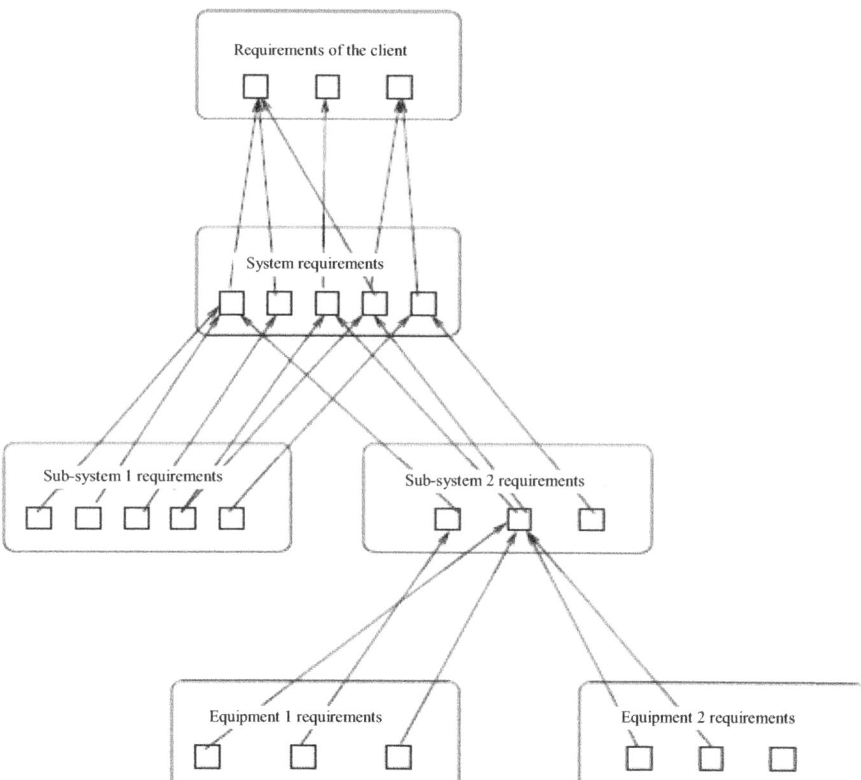

Figure 8.4. *Partial traceability between the requirements of the client and requirements linked to the equipment*

In [STAN 01], implementation of a requirements management environment is the best way to have a strong impact on the success of a project. Definition of a minimal set of requirements makes it possible to have a basis which can be managed, the tool then being a communication vector between teams.

Figure 8.4 shows how the recommendations of the client can be declined on the system and how processing can continue right down to the software and hardware-type elements. In the context of the verification phase of level n_i, it is necessary to show that the requirements of this level are in relation to the higher n_{i-1} level. This connection is made through the implementation of traceability. Implementation of traceability involves definition of at least one link between two objects. In the context of the requirements, traceability links must make it possible to show that an n_i-level requirement is linked to a need of the preceding n_{i-1} level. The inverse link makes it possible to show that no requirement has been forgotten during the production process.

As Figure 8.5 shows, there are currently several basis transformations of requirements. Among these, two cases are particularly interesting; the addition and deletion of a requirement; in either case, it is absolutely necessary to attach a justification to the requirement.

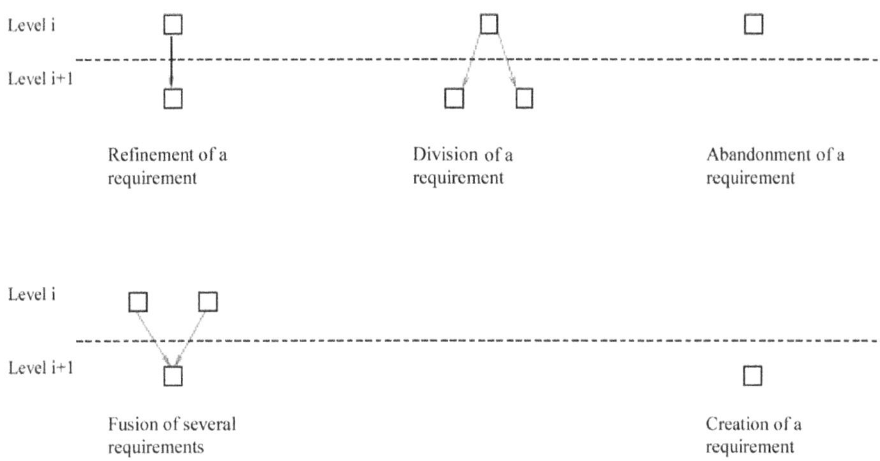

Figure 8.5. *Basic transformation of requirements*

8.2.4.2. *Specification*

The specification of a software application is thus at the very least, a set of requirements. A first analysis of the specifications provided by the client must make it possible to identify functional needs. Figure 8.6 presents the proposed process.

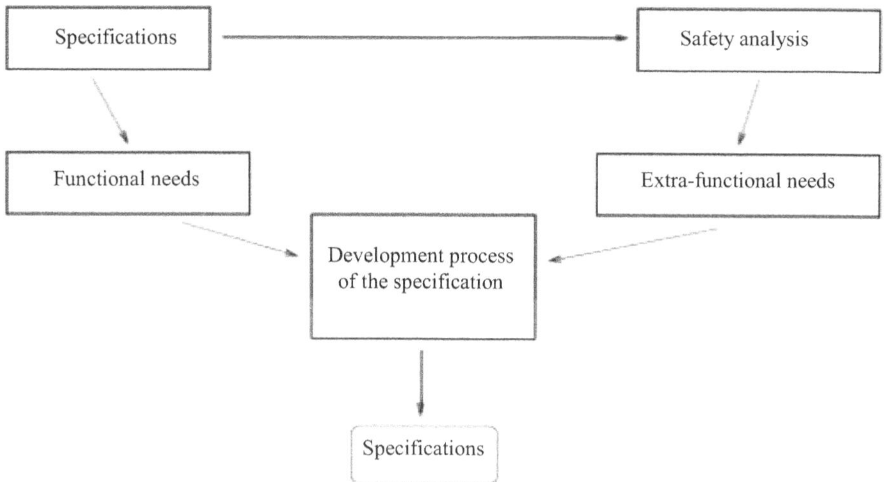

Figure 8.6. *Development of the specification*

In parallel with the identification of the functional needs, it is possible to begin analyses linked to dependability, the objective of which is to define the non-functional requirements: safety requirements but also availability, reliability, or other requirements. It is, however, necessary to know a bit more about the needs of the software application. Within the application, it is thus necessary to introduce:

– interfaces with the environment;

– the notion of state in establishing a partition between safe functioning, decline, and dangerous states;

– the notion of correct behavior and of dangerous behavior;

– the notion of requirement linked to safety; this type of requirement must allow for characterization of dangerous behavior;

– the integrity level of the software (written SSIL/DAL).

The environment of the software application is composed of interfaces with the hardware resources (memory, specific address, input/output, *watchdog*, etc.), with other software applications (base software, related application, etc.) and/or with the operating system. Figure 8.7 reveals a software application environment, which is composed of three entries (E_i), two exits (S_j), and three interfaces (I_k) with the hardware resources (example: access to a specific memory address).

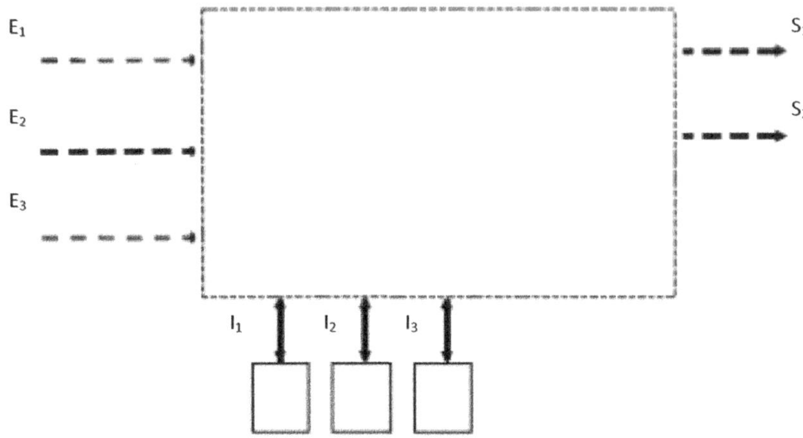

Figure 8.7. *Software application environment*

Standards recommend that the specification of a software application be composed from a textual description of the need (the requirements) and all the notations necessary to facilitate understanding of the need. So, for the EN 50128 [CEN 01] standard, there is Table A.2 (Table 8.2).

Classically, designers of a software application go directly from the textual requirements to the code without being able to verify coherence of the requirements and without consolidating the requirements.

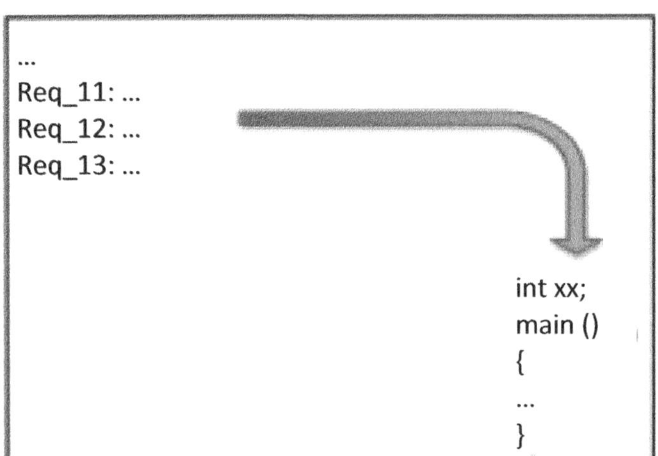

Figure 8.8. *From the requirements to the code*

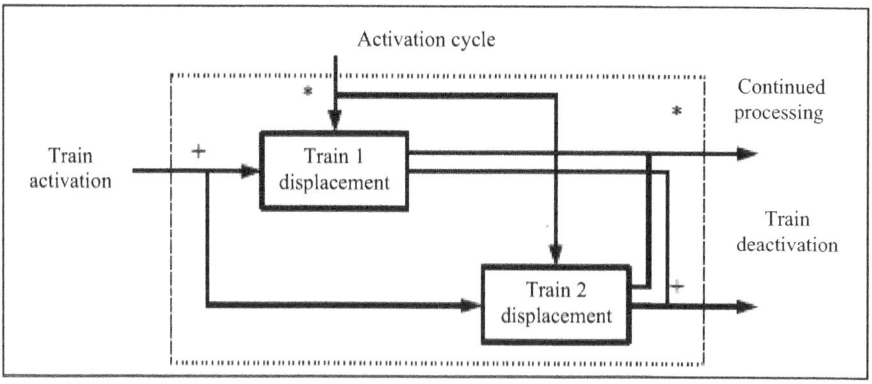

Figure 8.9. *Example of static introducing different types of communication*

Successful development of a maintainable software application involves at least two stages:

– a formalization of the requirements phase (see Figure 8.11). This formalization phase can rely on structured methods (for example, Figure 8.9 in SADT[12] makes it possible to establish a functional analysis, etc.), on semi-formal or formal methods (controller, Petri net, Grafcet, B method[13], SCADE[14], etc.);

– an architecture phase (see section 1.2.3.3).

The formalization phase is important because it enables translation of an abstract requirement into modeled elements, such as the following P1 property:

P1: "there must not be any risk of collision"

which can be translated into set-theoretic logic of the first order in:

$$\forall \{t_1,t_2\} \in [T], \textit{hence } D_{t_1} \cap D_{t_2} = \phi, \textit{ if } t_1 \neq t_2,$$

with D_{t1}, which is the domain of train speed i and $[T]$, which is the whole set of trains.

12 The acronym S.A.D.T stands for *Structured Analysis and Design Technique*. This method was invented by the Softech company in the US. The SADT method is a method of analysis by successive levels of a descriptive approach of a set, whatever it may be.

13 In Chapter 2 of [BOU 12b], we present an example of the use of SADT [LIS 90] and of the B method [ABR 96, BEH 96] in the context of the SAET-METEOR project [BOU 06, BOU 99].

14 Chapter 2 presents some examples of SCADE [DOR 08] use.

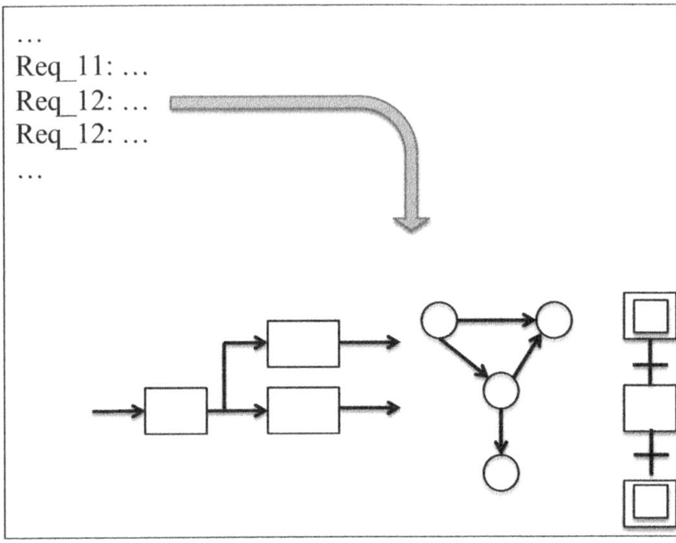

Figure 8.10. *Formalization of requirements*

8.2.4.3. *Synthetesis*

In the railway sector, structured method (SADT and SART for example), semi-formal method, and formal method (B method, SCADE, etc.) are used to formalize the need (set of requirement). More generally, we used a model.

The model provides the capability:

– to manage the need;

– to formalize the need to verify the coherence and the completeness;

– to help for test case selection.

	SSIL0	**SSIL1**	**SSIL2**	**SSIL3**	**SSIL4**
Formal method (based on a mathematic approach)	-	R	R	HR	HR
Modelisation	R	R	R	HR	HR
Structure of methodology	R	R	R	HR	HR
Decision table	R	R	R	HR	HR

Table 8.4. *New EN 50128 – Table A.2*

In the new version of the CENELEC EN 50128 (see Table 8.4), the notion of semi-formal (see Table 8.2) disappears and is replaced by modelization.

8.2.4.4. *Model*

The new version of the EN 50128 [CEN 11] standard (Table A.3) recommends that software application architecture is based on a structured method (SADT, etc.).

Yet it is possible to establish a model, which is based on the techniques of Table A.17 (see Table 8.5).

	SSIL0	SSIL1	SSIL2	SSIL3	SSIL4
Data modeling	R	R	R	HR	HR
Data organization	-	R	R	HR	HR
Command data flow	R	R	T	HR	HR
Finite state controllers/transition schematics	-	HR	HR	HR	HR
...					
Formal methods	R	R	R	HR	HR
Sequencing diagrams	R	HR	HR	HR	HR

Table 8.5. *New CENELEC EN 50128 [CEN 01] – Table A.17*

The creation of a model M is to help to understand a problem or a situation. In general, the specification phase, which allows for adaptation of the specifications, involves creation of an M model.

A model may be more or less close to the system studied, which would then be termed abstraction. The closer the model, the closer are the results obtained to those observed on the final system.

Another characteristic of models is their dependence on a semantic for language support.

The presence of a semantic enables implementation of reasoning techniques, which guarantee correction of obtained results.

A model (Figure 8.11) is in general composed of two complementary parts:

– a static model (Figure 8.9) describing the entities constituting the system and the states, which are able to be associated with them;

– a dynamic model describing authorized changes of state.

Figure 8.11 presents an example of a UML model[15] [OMG 06a, OMG 06b, OMG 07] of a rail system implementing different models: use case, sequence diagram, state/transition diagram, and class diagram.

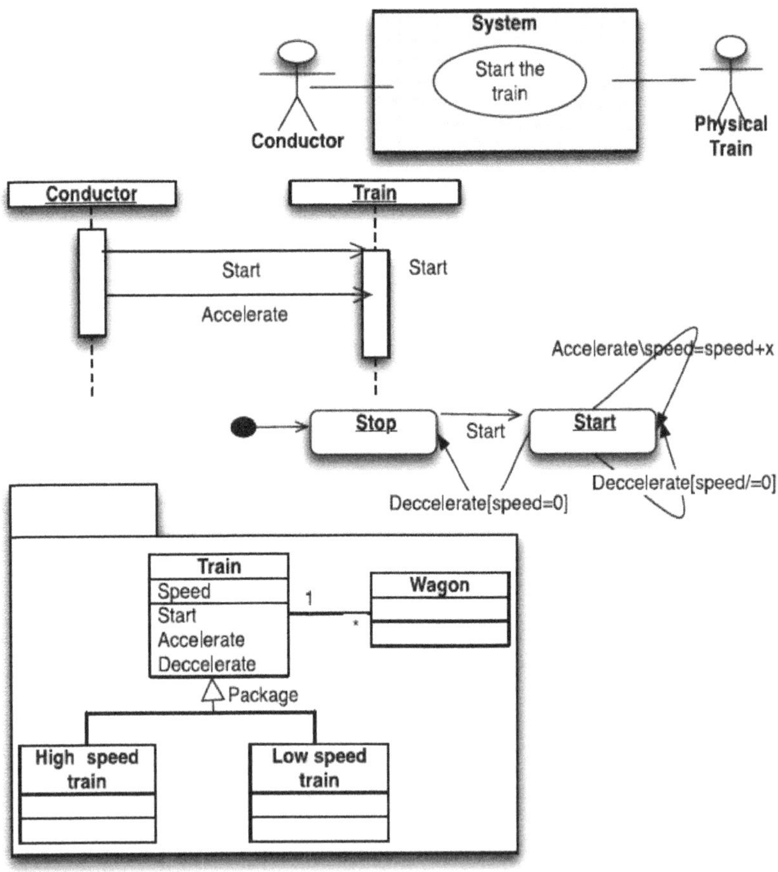

Figure 8.11. *Example of a UML model*

15 To find out more, visit: the OMG site: www.omg.org.

Using UML notation is thus not without certain questions [BOU 07, OKA 07]. How to use a notation without a semantic? How to evaluate an application based on UM notation? etc. Several studies aim to offer answers to these questions, for example [OOT 04, MOT 05].

At present, UML notation [OMG 06a, OMG 06b, OMG 07] is not recognized as a structured and/or semi-formal method in the standards, even if many would like to implement it completely or partially.

In [BOU 07, BOU 09a], and Chapter 19 of [RAM 09], we have shown how UML notation can be used to create some models of safety-critical systems.

In the context of the RT3-TUCS, ANR-SAFECODE, and ANR-RNTL-MEMVATEX projects, we have studied different ways of introducing UML as a means of modeling a safety-critical system; see for example [RAS 08, BOU 08, IDA 07a, IDA 07b, OKA 07, BOU 05].

Over the last few years, several studies have enabled a formalization of UML notation to be proposed through returning to formal languages, for example [IDA 09, IDA 06, MAR 04, MAR 01, MAM 02, LED 02]. Studies such as [MOT 05] must also be noted, which have made it possible to suggest some additional rules for verifying UML models.

8.2.4.5. *Architecture*

Once the software application specification is created, it is then possible to establish an architecture. This architecture aims to break down the software application into components[16] (or modules, following the vocabulary used).

Figure 8.12 presents an example of architecture. The software application has been broken down into three components (C_1, C_2, and C_3); there are interfaces between the environment (E_1, E_2, S_1 and S_2) and components and some interfaces between components; for each component, the list of requirements to take into account is given.

Creation of a software application specification and/or an architecture (though the same could be said of the design phase) must be able to continue processing (Figure 8.13) of the specification phase up to the design phase; not to mention the coding phase.

---

16 In the new versions of the standards, the notion of components is used whereas in preceding versions, "module" (file, functions, class, etc.) was used.

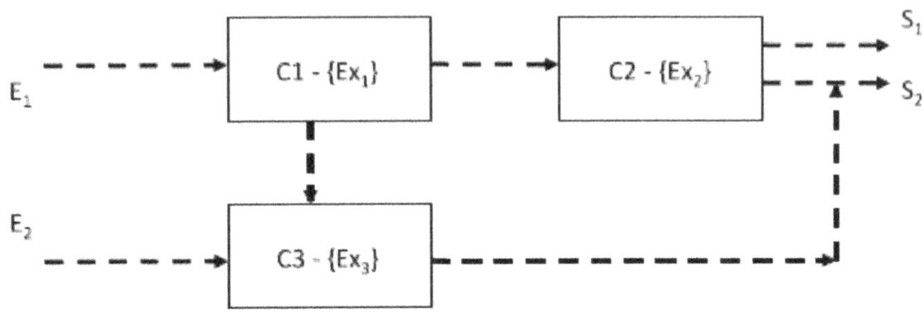

Figure 8.12. *Example of a software application architecture*

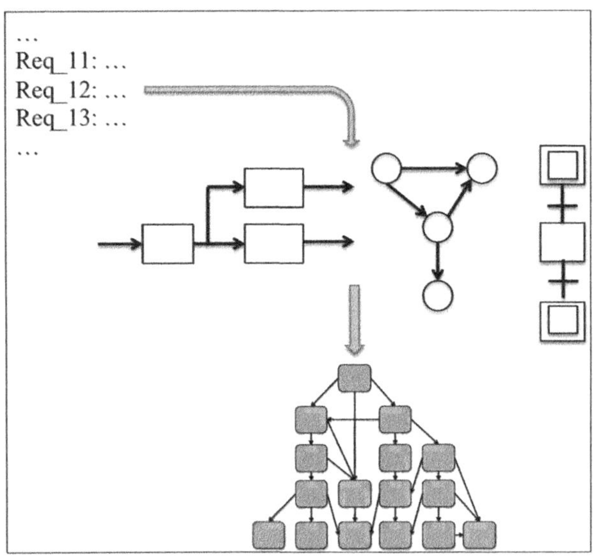

Figure 8.13. *From the requirements to the architecture*

8.2.5. *Coding*

8.2.5.1. *Presentation*

The classic development process of a software application is based on the use of a programming language, for example Ada [ANS 83, ISO 95] C [ISO 99], and/or C++ [ISO 03].

Even if these languages have a certain abstraction level with respect to the code ultimately executed on the computer, they necessitate the writing of a line of code.

	SSIL0	SSIL1	SSIL2	SSIL3	SSIL4
Ada	R	HR	HR	R	R
MODULA-2	R	HR	HR	R	R
PASCAL	R	HR	HR	R	R
C or C++ without restriction	R	-	-	NR	NR
Sub-assembly of C or C++	R	R	R	R	R

Table 8.6. *CENELEC EN 50128 [CEN 01] – Table A.15*[17]

It is not possible to analyze all the current programming languages used for all industries. In the rest of this section, we will analyze the advancement, which has taken place in the rail sector.

8.2.5.2. *Assessment*

8.2.5.2.1. The Ada language

The first rail applications in France were programmed, in the middle of the '80s, with the Modula 2 language. Since then, however, the Ada 83 language [ANS 83] has become the reference language for the development of safety-critical applications [RIC 94].

```
with Ada.Text_IO;
procedure Hello is
begin Ada.Text_IO.Put_Line("Hello, world!");
end Hello;
```

Figure 8.14. *Example of Ada code*

17 The CENELEC [CEN 01, CEN 00, CEN 03] referential, like the CEI/IEC 61508 [IEC 98] standard, introduces the notion of SIL for *Safety Integrity Level,* which can take four values, from 1 to 4. To find out more on SIL management, we advise you to read Chapter 7 of [BOU 11]. The notion of SSIL (*Software* SIL) concerns assignment of a safety level to the software aspect. The SSIL level relates to systematic failures. SSILs can go from 0 up to 4. SSIL0 software is software which has no impact on safety and must, at the very least, implement the requirements of the ISO 9001:2008 [ISO 08].

As Table 1.1 shows, in the context of applications, which have a high level of criticality (SSIL3/SSIL4), the Ada language itself is only R (recommended); it is necessary to establish a sub-assembly of the language so that the use of Ada is HR (highly recommended).

Ada was conceived as an initiative of the DoD (the US Department of Defense) to federate more than 400 languages or dialects used by this body since the 1970s.

Ada is very widely used in the framework of embedded software applications in avionics (airbus), in space (the Ariane rocket), and in the rail sector. The principal characteristic of these systems is that they require a correction of the execution.

The Ada 83 language [ANS 83] has advanced toward a second major standard, Ada 95 [ISO 95], which is the first object-normalized language. It provides the possibility of constructing object-oriented models. The latest version to date is Ada 2005.

DEFINITION 8.5.– CERTIFICATION. Certification consists of achieving a certificate, which is a commitment that the product complies with a normative referential. Certification is based on the results of an evaluation and on production of a certificate.

Regarding certification of Ada compilers, the existence of a standard and of a rather fine semantic of Ada has made it possible to define a compiler certification process. This process has been implemented on different compilers and is based on a suite of tests named ACATS (*Ada Conformity Assessment Test Suite*); see the [ISO 99] standard. To find out more, consult [ADA 01].

For the moment, these new versions of Ada have not been adopted by the embedded systems sector because of this object-oriented aspect. Given their efficacy, however, Ada 95 compilers are used for compilation provided that a sub-assembly of the language, which does not use the "object-oriented" features is relied upon.

John Barnes, in the context of the ADACORE article makes a presentation on the strengths of the Ada 2005 language (syntax and semantic, strong typing, sound management of the pointer and the memory, etc.) and the impact of its implementation, upon demonstration that the software is reliable.

The "object-oriented" aspect is not taken into account by CENELEC EN 50128 ([CEN 01], DO 178 [ARI 92], CEI/IEC 61508 [IEC 98], and ISO 26262 [ISO 09]) standards applicable to safety-critical applications.

To get around this stumbling block, the ISO 15942 [ISO 00] standard delineates a restriction on Ada 95 language constructions and defines some rules of use (programming style), which enable creation of a supposedly-certifiable application (see definition 1.4).

DEFINITION 8.6.– CERTIFIABLE APPLICATION. A certifiable software application is a software application, which has been created in order to be certified.

The SPARK Ada language [BAR 03] is a programming language, which is a sub-assembly of Ada. All the complex structures of Ada considered as risky or not allowing for an easy safety demonstration are omitted in SPARK Ada. A mechanism enabling the addition of annotations in the code has been introduced.

SPARK Ada tools contain a compiler, but also a verifier for annotations. It is to be noted that there is a free version of the SPARK Ada tools.[18] Chapter 1 of this book provides the opportunity to present the SPARK Ada tool as well as some industrial examples of its implementation.

8.2.5.2.2. The C language

The C[19] language [KER 88] is one of the first languages to be made available to developers to create complex applications. The principal difficulty of C resides in the partial definition of the language, which means that different compilers generate one file with different behaviors. It has since been subjected to a standardization process by the ANSI [ISO 99].

	SSIL0[4]	SSIL1	SSIL2	SSIL3	SSIL4
Ada	R	HR	HR	HR	HR
MODULA-2	R	HR	HR	HR	HR
PASCAL	R	HR	HR	HR	HR
C or C++ without restriction	R	R	R	R	R
C#	R	R	R	R	R
JAVA	R	R	R	R	R

Table 8.7. *New CENELEC EN 50128 [CEN 11] – Table A.15 (partial)*

18 To find out more about AdaCore and free tools, such as GNAT and SPARK Ada, visit: www.libre.adacore.com.

19 Although "Kerdigan and Ritchie" [KER 88] does not follow the ANSI on the C language, it remains one of the most interesting books on the subject.

Regarding use of the C language [ISO 99], contingent upon the safety level required, the CENELEC EN 50128 [CEN 01] standard recommends defining a sub-assembly of the language (see Table 8.7), the execution of which is controllable.

Table 8.7 (an extract from the new version of the CENELEC EN 50128 [CEN 11] standard) shows that there was sufficient experience feedback for the Ada, C, and C++ languages, enabling no further explicit mention of the notion of sub-assembly of the language as it is taken as established.

Figure 8.15 presents a piece of C code, which can give two different codes contingent upon anomaly (a) or (b), which is implemented. This example emphasizes the weaknesses of the C language and small programming errors, which are not detected in the compilation. This type of error is detected if the programming language used is Ada.

Being that the C fragment of program is as follows

```
if (TheSignal == clear)
{
        open_gates();
        Start_train();
}
```

(a) Failure →

```
if (TheSignal == clear) ;
{
        open_gates();
        Start_train();
}
```

(b) Failure ↓

```
if (TheSignal = clear)
{
        open_gates();
        Start_train();
}
```

Figure 8.15. *Example*[20] *of a fault in C*

Certain weaknesses of C may be overcome by implementing some programming rules; by way of an example, to avoid an anomaly of the type if (a = cond) instead of if (a == cond), a rule of the following form can be implemented: "in the framework of comparison with a variable, this must be in the left part of the expression".

20 This example is retrieved from ADACORE.

From 1994, some experience feedback regarding the implementation of C (see for example [HAT 94]) has revealed that it was possible to define a sub-assembly of C usable to create software applications needing a high level of safety (SSIL3 and SSIL4).

In fact, since the end of the 1990s, the MISRA-C [MIS 98, MIS 04] standard, which was developed by the *Motor Industry Software Reliability Association* (MISRA[21]) has become a standard for the C language.

MISRA-C [MIS 04] specifies some programming rules (see the examples of Table 8.8) making it possible to avoid execution errors provoked by poorly-defined constructions, unforeseen behaviors (certain C language structures are not completely defined) and misunderstandings between those people in charge of production (legible code, implicit code, etc.). Several tools enable the MISRA-C rules to be automatically verified.

ID	Status[22]	Description
Rule 1.1	Required	All of the code must conform to the ISO 9899:1990 norm "Programming languages – C", amended and corrected by ISO/IEC9899/CORI:1995, ISO/IEC9899/AMD1:1995 and ISO/IEC9899/COR2:1996
Rule 5.4	Required	Each tag is a single identifier
Rule 14.1	Required	There must not be any dead code
Rule 14.4	Required	No unconditional jumps (goto) in programs
Rule 14.7	Required	A function must have a single output point at the end of the function
Rule 17.1	Required	The pointer arithmetic can only be used for pointers which address a table or tabular element
Rule 17.5	Advisory	An object declaration must not contain more than two levels of pointer indirection

Table 8.8. *Some MISRA-C: 2004 [MIS 04] rules*[23]

21 To find out more, visit: www.misra.org.uk.

22 A MISRA rule may be "required" or "advisory". A "required" rule must be compulsorily implemented by the developer and an "advisory" rule cannot be disregarded, even if it is not compulsory to implement it.

23 [MIS 04] introduces 122 "required" and 20 "advisory" rules.

The MISRA-C [MIS 04] standard repeats some rules, which are explicit in several standards (see for example 14.4 and 14.7):

– rule 14.4: in the EN 50128 standard – Table A.12 or the IEC 61508 standard – Table B.1;

– rule 14.7: in the EN 50128 standard – Table A.20 or the IEC 61508 standard – Table B.9;

MISRA-C [MIS 98] was created in 1998 and updated in 2004 [MIS 04], which makes use of experience feedback.

The principal difficulty of C remains the choice of a compiler having sufficient experience feedback at its disposal for the chosen target and safety level to be achieved. In the absence of a precise and complete standard, there is currently no certification process for C compilers, even though there are initiatives such as [LER 09].[24]

8.2.5.2.3. Object-oriented language

As already stated, the "object-oriented" aspect is not taken into account by the CENELEC EN 50128 ([CEN 01] and DO 178 [ARI 92], CEI/IEC 61508 [IEC 98] and ISO 26262 [ISO 09]) standards are applicable to safety-critical applications.

	SSIL0	**SSIL1**	**SSIL2**	**SSIL3**	**SSIL4**
No dynamic objects	-	R	R	HR	HR
No dynamic variables	-	R	R	HR	HR
Limited usage of pointers	-	R	R	R	R

Table 8.9. *CENELEC EN 50128 [CEN 01] – Table A.12*

The "object-oriented" aspect is cited in the CENELEC EN 50128 [CEN 01] standard, but the constraints applying to languages do not allow for development of a safety-critical application (SSIL3 and SSIL4) with this type of language (see Table 8.9).

24 It is to be noted that [LER 09] features a verification job by a limited C compiler, but that if the term "certification" is to be introduced, it does not concern certification by an external body. Certification does not only involve demonstration that the compiler is correct but must also apply to the process implemented, to the elements produced (documents, sources, etc.) and to the tools used (Ocaml, Coq, and other tools must be demonstrated, such as safety).

As shown by Tables 8.6 and 8.7, the C++ language [ISO 03] is cited as applicable but certain recommendations are not compatible with the use of an object-oriented language, as shown in Table 8.9.

C++ was developed during the 1980s; it proceeded from an improvement of the C language. C++ introduces the notion of class, inheritance, virtual functions, and overload. It was standardized by the ISO in 1998 and in 2003 [ISO 03].

ID	Status[25]	Description
Rule 0-1-1	*Required*	A software must not contain non-achievable code.
Rule 0-1-2	Required	A software must not contain non-executable paths.
Rule 1-0-1	Required	All the code must conform to ISO/IEC 14882:2003 "The C++ *standard incorporating Technical Corrigendum* 1".
Rule 2-10-4	Required	A "class", "union" or "enum" denomination must be a single identifier.
Rule 5-2-3	Advisory	Operation of "cast" of a base class towards a derived class must not be made using polymorphous types.
Rule 15-5-1	Required	A class destructor must not quit with an exception.
Rule 17-0-4	Docuement	All libraries must conform to MISRA-C++.
Rule 18-0-1	Required	The C library must not be used.

Table 8.10. *Some MISRA-C++:2008 [MIS 08] rules*[26]

Since the beginning of the 2000s, several projects have been undertaken to define a framework enabling the use of the C++ language for the development of high-safety level applications (SSIL3, SSIL4).

25 A MISRA rule may be "required", "advisory", or "document". A "required" rule must compulsorily be implemented by the developer, an "advisory" rule cannot be disregarded, even if it is not compulsory to implement it and a "document" rule is compulsory.
26 [MIS 08] introduces 198 rules.

The following studies can be cited:

– JSF++ (*Joint Strike Fighter* C++), which has published a guide [LOC 05] regarding the current work, notably the MISRA-C: 1998 [MIS 98] standard;

– MISRA, which has developed the MISRA-C++:2008 [MIS 08] standard; Table 1.5 shows some MISRA-C++:2008 rule examples;

– OOTIA[27] (*Object Oriented Technology in Aviation*), which has published several guides [OOT 04a, OOT 04b, OOT 04c], and [OOT 04].

C++ [ISO 03] is therefore quite an old language. Some approaches identifying the weak points of C++ and proposing some rules appeared quite early on [MEY 98, SUT 05], but the definition of a framework for using C++ for high safety-level applications is quite recent [OOT 04, LM 05, MIS 08], which explains how applications in C++ are to be found even at SSIL2.

As Table 8.10 shows, the MISRA C++: 2008 [MIS 08] standard introduces some rules inspired by those current for the C language, but which do not allow for all the difficulties of C++ to be taken into account.

Under various pressures (a reduction in the number of Ada and C programmers, for example), the updating of standards, such as CENELEC EN 50128 or DO 178 has given rise to initiatives seeking to introduce object-oriented languages.

	SSIL0	SSIL1	SSIL2	SSIL3	SSIL4
The classes should only have a single objective.	R	R	R	HR	HR
Inheritance used only if the derived class is a refinement of its base class.	R	HR	HR	HR	HR
Depth of inheritance limited by coding norms.	R	R	R	HR	HR
Neutralization of operations (methods) under strict control.	R	R	R	HR	HR
Multiple inheritance used only for interface classes.	R	HR	HR	HR	HR
Inheritance from unknown classes.	-	-	-	NR	NR

Table 8.11. *New CENELEC EN 50128 [CEN 11] – Table A.23*

27 To find out more, visit: www.faa.gov/aircraft/air_cert/design_approvals/air_software/oot.

Thus, the new version of the CENELEC EN 50128 [CEN 11] standard has expanded the list of usable object-oriented languages to JAVA and to C#, as Table 8.11 shows. Yet, this new version of the standard introduces some restrictions (limitations on inheritance) as Table 8.6 shows.

Figure 8.16. *Example[28] of an actual programmable controller*

The C version of the DO 178 standard should use a specific annex aiming to define the constraints of implementing an object-oriented language for production of a safety-critical application.

As for C, the difficulty with C++ remains demonstrating that the compiler for the chosen target and associated libraries complies with the safety objectives, which have been defined by safety studies. As there is currently no certification for C++ compilers, it will be necessary to establish a justification based on experience feedback and/or qualification.

28 Photograph taken by Jean-Louis BOULANGER.

8.2.5.3. *Programmable controller*

The "continuous processes" CEI/IEC 61511 [IEC 05] industry standard brings precision in restricting the initial field of application of the CEI/IEC 61508[29] [IEC 98] standard to the traditional context of continuous processes.

The CEI/IEC 61511 [IEC 05] standard applies in the framework of E/E/PE systems used in the context of safety instrumented functions for controlling a process (chemical products, petrol refining, petrol and gas production, non-nuclear electricity production). It does not apply to manufacturers or system providers, who must use the CEI/IEC 61508 standard.

For synthesizing, the CEI/IEC 61511 standard is used when already-certified components (of programmable controllers) are used and when they are programmed with limited variability languages.

With respect to programming languages, the CEI/IEC 61511 [IEC 05] standard only concerns programming languages common to the industry, which are specified through the CEI/IEC 61131-3 [IEC 03] standard.

This in no way prevents anyone from programming their E/E/PE system with the Ada or C++ languages, but in these cases, it is necessary to refer to the CEI/IEC 61508 [IEC 98] standard, which defines how these languages can be used.

Part 3 of the CEI/IEC 61131-3 [IEC 03] standard describes the principles for producing an application for a programmable controller and introduces several technical terms: FBDs for functional block diagrams, SFCs for sequential function charts, suites of instructions (written IL), structured text (ST), and contact diagrams (LD for ladder diagrams).

The languages introduced by the CEI/IEC 61131-3 [IEC 03] standard are used more and more for making software applications. This is because these graphic languages are easily assimilated and standardized. There are thus some tools that cover all the formalisms described in the CEI/IEC 61131-3 [IEC 03] standard.

These tools are supplied for making programmable controller applications and may thus come with a certificate for a safety level, which might be SSIL2 or SSIL3.

The ControlBuild tool is presented in Chapter 8 of [BOU 12c]. ControlBuild is used by several actors in the rail sector to make SSIL2-level safety-critical software applications.

---

29 In Chapter 10 of [BOU 09b], there is a discussion on the use of SISs (Safety Instrumented Systems) in industry. Chapter 6 of [BOU 11] presents a complete example of implementation of a programmable safety controller for managing ventilation of the Soumagne tunnel and the Walhorn trench of the high-speed L3 line.

8.2.5.4. *Qualification*

The different standards (DO 178, ISO 26262, CENELEC EN 50128) reveal/or will reveal the notion of the "qualification file".

Qualification of a tool depends on its impact on the final product:

– an impact on the executable (compiler, data generator, etc.);

– an impact on the verification and/or validation (test tools, etc.);

– no impact.

The qualification file can rely on different activity types:

– analysis of the use of a certificate (commonly termed cross-acceptance): if a certificate exists, it is necessary to verify that the tool is suited to what it is intended to be used for (same normative referential, similar safety level, etc.);

– experience feedback construction: an inventory of all uses (those that are known and those of the business). It must then be capable of justifying the version used, levels of safety, size, sectors, etc.;

– establishment of a qualification; test runs of the tool can be undertaken so as to show that the former is usable for a given safety level.

The notion of qualification is an important notion, which is linked to the implementation of tools in the production process of a software application.

All the tools used can be subjected to a qualification phase.

Regarding formal methods, the question of qualification of tools is of fundamental importance, since the complexity of technologies implemented (prover, model-checking, etc.), the confidentiality aspect (licensed algorithm, etc.), the innovation aspect (new technology, few users, etc.), and the maturity aspect (product resulting from research, product made using a "free" license, etc.) do not permit trust to be built easily.

8.2.5.5. *Advantages/disadvantages*

The programming language used must facilitate anomaly detection and for this, it is necessary to implement strong typing and modular and structured programming.

As Table 8.12 shows, one of the principal difficulties while using a language such as C or C++ resides within demonstrating the "proven" character of the translator. Use of a programming language:

– necessitates definition of a subset of the language (limiting the difficulties) and using a set of design and coding rules;

– necessitates the use of sufficient experience feedback to show that the tools and the compiler are usable for the expected safety level.

	SSIL1	SSIL2	SSIL3	SSIL4
Adequate programming language	HR	HR	HR	HR
Strongly typed programming language	HR	HR	HR	HR
Sub-assembly of the language	-	-	HR	HR
Certified tools	R	HR	HR	HR
Tools: confidence in which has increased as a result of use	HR	HR	HR	HR
Certified translator	R	HR	HR	HR
Translator: confidence has been increased as a result of use	HR	HR	HR	HR
Library of modules software and proven/verified components	R	HR	HR	HR

Table 8.12. *IEC 61508 [IEC 98, part 3] – Table A.3*

8.2.6. *Verification and validation (V&V)*

Creation of a software application must take into account not only the design of the software application, but also activities that allow for it to reach a certain quality standard.

Development of a certifiable software application is constrained by the requirements of standards associated with each industry (aeronautical [ARI 92], automobile [ISO 09], rail [CEN 01, CEN 11], nuclear [IEC 06], and generic [IEC 98]).

Accomplishment of a quality standard requires demonstration that no fault has been introduced during the design and that the product corresponds to the needs, which have been identified. Above all else, it is necessary for us to recall what verification and validation are.

These normative requirements recommend implementation of a "V-cycle"-type development process based on the activities of verification and validation, based on the carrying out of tests (MT, IT and FT[30]).

Activity testing implementation suffers from several problems, which include:

– the cost and complexity of activity testing;

– late detection of faults;

– the difficulty of carrying out all the tests.

This is why it is necessary to implement other practices, which must enable faults in the software application to be detected as early as possible.

One of the possible orientations consists of implementing formal methods (for example, the B method [ABR 96], SCADE[31] [DOR 08], VDM [JON 90], Z [SPI 89], etc.), which on the basis of a model and set of properties, enable demonstration that the software product verifies the stated properties.

It may, however, be interesting, on the basis of classic development languages (such as C) to explore the program behaviors and to demonstrate that it verifies some properties.

	SSIL0	SSIL1	SSIL2	SSIL3	SSIL4
Formal proof	-	R	R	HR	HR
Static analysis	-	HR	HR	HR	HR
Dynamic analysis and tests	-	HR	HR	HR	HR
Metrics	-	R	R	R	R
Tracability	R	HR	HR	M	M
AEEL	-	R	R	HR	HR
Coverage of the test for the code	R	HR	HR	HR	HR
...					

Table 8.13. *New CENELEC EN 50128 ([CEN 11]) – Table A.5*

30 MT for module testing, IT for integration testing and FT for functionnal testing.
31 SCADE is distributed by Esterel Technologies; see: www.esterel-technologies.com.

This analysis of behaviors can be made through implementation of static analysis techniques of code based on abstract interpretation [COU 00] and proof of programs [HOA 69]. The first volume of this series [BOU 11a] is dedicated to abstract interpretation.

It should be noted that one of the difficulties in implementing these techniques lies in the absence of recognition of these techniques within the current standards. In effect, certain standards (see Table 8.4 and 8.13, extract from the standard CENELEC EN 50128 [CEN 01a] for example) advocate implementation of formal methods, but they do not mention the notion of abstract interpretation (or derived methods).

In the new version of the CENELEC EN 50128, (see Table 8.13) static analysis and the formal proof are directly cited as such verification techniques. In Table A.13, we also see the notion of traceability that is very important for demonstrating the safety of the application.

8.3. Conclusion

This chapter presents some requirements to respect when a software application with a high safety objective (SSIL3/SSIL4) is to be created. In the outlines of the automobile [ISO 09], rail [CEN 01], and generic [IEC 98] industry standards, formal methods are advocated (see Tables 8.4 and 8.13) as a means of satisfying needs and achieving a controlled code production.

In effect, some formal methods[32], such as the B method [ABR 96], SCADE [DOR 08], VDM [JON 90], Z [SPI 89], etc. give you the possibility to acquire the user need, to formalize as a model, to demonstrate that some properties are respected and to generate some code that are the image of the model (no verification is needed on the code if you are confident on the safety of your tools).

This book and [BOU 12a, BOU 12b] cover different aspects of formal techniques: static analysis of code, formal methods, and tools.

The formal techniques and methods are now successfully being used for projects of various sizes in industry. The associated tools have reached maturity so that we can account for the complexity of such applications. Note that the complexity of industrial applications very often has an impact on the processing time[33].

---

32 The list is not meant to be total; for more information, see: for example [MON 00].

33 Within the context of the SAET-METEOR, [MAT98] it took longer than a week in 1998 to analyze 100,000 lines of Ada code with the Polyspace® tool (see Chapter 8 for more information), whereas maintaining that would take one or two hours.

Taking the techniques and formal methods that have an impact on the implemented process into account, it is necessary to build a new standard that conforms to the quality standards being enforced. This standard does not take into account that the model can replace some documents. It is necessary to explain how the process based on a model gives you the same level of confidence on the safety of the software. The confidence is based on the global process and on the effectiveness of the verification, actually using a model does not provide the same level of verification, in fact some very important verifications are removed and not replaced by equivalent activities. Indeed, if the tool is no longer maintained and formalism is proprietary, it is difficult to update the application without the model.

Another difficulty of the *model-centered* approach is bringing the software application's level of safety to that of the tools implemented.

Though it is difficult to show that a compiler C is SSIL4, it is even more difficult to show that a prover is SSIL4. Within the context of compilers, it is possible to set up strategies based on redundancy and diversity. As for specific tools such as provers, *model-checking* tools, and/or tools for abstract interpretation, it is difficult to have two tools of similar effectiveness within the same field[34]. Therefore, it is necessary to set up tool qualification files.

The different standards (DO 178, ISO 26262, CEI/IEC 61508, CENELEC EN 50128) show the concept of a *qualification file*. The qualification of a tool depends on its impact on the final product.

Concerning formal techniques and methods, the question of tool qualification is of primary importance because the complexity of the implemented technologies (prover, *model-checking*, etc.), the confidentiality aspect (licensed algorithm, etc.), the innovation aspect (new technology, few users, etc.), and the maturity aspect (product resulting from research, carried out under a "free" license, etc.) do not bring confidence on the safety.

One of the difficulties in implementing these techniques lies in the absence of recognition of these techniques within the current standards. In effect, certain standards (see Table 8.13, extract from the CENELEC EN 50128 [CEN 01] standard for example) advocate implementation of formal methods but they do not mention the notion of abstract interpretation (or derived methods).

34 Recall that for several of the tools currently in use, the algorithms are not public and are also under *copyright*.

8.4. Bibliography

[ABR 96] ABRIAL J.R., *The B-Book - Assigning Programs to Meanings*, Cambridge University Press, Cambridge,1996.

[ADA 01] ADA RESOURCE ASSOCIATION, Operating Procedures for Ada Conformity Assessments, version 3.0, www.ada-auth.org/procs/3.0/ACAP30.pdf, April 2001.

[ANS 83] ANSI, Norme ANSI/MIL-STD-1815A-1983, Langage de programmation Ada, 1983.

[ARA 97] ARAGO, "Applications des méthodes formelles au logiciel", *Observatoire français des techniques avancées* (OFTA), vol. 20, Masson, Paris, June 1997.

[ARI 92] ARINC, Software Considerations in Airborne Systems and Equipment Certification, DO 178B and EUROCAE, no. ED12, edition B, 1992.

[BAI 08] BAIER C., KATOEN J.-P., *Principles of Model Checking*, The MIT Press, Cambridge, MA, 2008.

[BAR 03] BARNES J., *High Integrity Software: The SPARK Approach to Safety and Security*, Addison Wesley, Boston, 2003.

[BEH 93] BEHM P., "Application d'une méthode formelle aux logiciels sécuritaires ferroviaires", *Atelier Logiciel Temps Réel, 6e Journées internationales du Génie Logiciel*, 1993.

[BEH 96] BEHM P., "Développement formel des logiciels sécuritaires de METEOR", dans H. Habrias (ed.) *Proceedings of 1st Conference on the B method, Putting into Practice methods and tools for information system design*, p. 3-10, IRIN Institut de recherche en informatique, Nantes, November 1996.

[BEN 03] BENVENISTE A., CASPI P., EDWARDS S.A., HALBWACHS N., LE GUERNIC P., DE SIMONE R., "The synchronous languages 12 years later", *Proceedings of the IEEE*, vol. 91, no. 1, January 2003.

[BOU 99] BOULANGER J.-L., DELEBARRE V., NATKIN S., "METEOR: validation de spécification par modèle formel", *Revue RTS*, no. 63, p. 47-62, April-June 1999.

[BOU 05] BOULANGER J.-L., BERKANI K., "UML et la certification d'application", *ICSSEA 2005*, CNAM, Paris, 1-2 December 2005.

[BOU 06] BOULANGER J.-L., Expression et validation des propriétés de sécurité logique et physique pour les systèmes informatiques critiques, thesis, Compiègne University of Technology, 2006.

[BOU 07] BOULANGER J.-L., BON P., "BRAIL: d'UML à la méthode B pour modéliser un passage à niveau", *Revue RTS*, no. 95, p. 147-172, April-June 2007.

[BOU 07a] BOULANGER J.-L., "UML et les applications critiques", *Proceedings of Qualita '07*, p. 739-745, Tangiers, Morocco, 2007.

[BOU 08] BOULANGER J.-L., "RT3-TUCS: How to build a certifiable and safety critical railway application", *17th International Conference on Softwaree Engineering and Data Engineering, SEDE-2008*, p. 182-187, Los Angeles, 30 June-2 July 2008.

[BOU 09a] BOULANGER J.-L., IDANI A., PHILIPPE L., "Linking paradigms in safety critical systems", *Revue ICSA*, September 2009.

[BOU 11] BOULANGER J.-L. (ed.), *Sécurisation des architectures informatiques industrielles*, Hermes Lavoisier, Paris, 2011.

[BOU 11a] BOULANGER J.-L. (ed.), *Utilisation industrielles des techniques formelles – interprétation abstraite*, Hermes Lavoisier, Paris, 2011.

[BOU 11b] BOULANGER J.-L. (ed.), *Techniques industrielles de modélisation formelle pour le transport*, Hermes Lavoisier, Paris, 2011.

[BOU 12a] BOULANGER J.-L. (ed.), *Static Analysis of Software - The Abstract Interpretation*, ISTE,London, John Wiley & Sons, New York, 2012.

[BOU 12b] BOULANGER J.-L. (ed.), *Formal Methods*, ISTE,London, John Wiley & Sons, New York, 2012.

[BOU 12c] BOULANGER J.-L. (ed.), *Outil de mise en oeuvre industrielle des techniques formelles*, Hermes Lavoisier, Paris, 2012.

[BOU 12d] BOULANGER J.-L. (ed.), *Mise en œuvre de la méthode B*, Hermes Lavoisier, Paris, forthcoming 2012.

[BOW 95] BOWEN J.P., HINCHEY M.G., *Applications of Formal Methods*, Prentice Hall, 1995.

[CEN 00] CENELEC – EN 50126. Applications Ferroviaires. Spécification et démonstration de la fiabilité, de la disponibilité, de la maintenabilité et de la sécurité (FMDS), January 2000.

[CEN 01a] CENELEC – EN 50128. Railway applications, Communications, signalling and processing systems, Software for railway control and protection systems, May 2001.

[CEN 01b] CENELEC - EN 50159-1 Railway applications, Communication, signalling and processing systems, Safety related communication in closed transmission systems, 2001.

[CEN 01c] CENELEC - EN 50159-2 Railway applications, Communication, signalling and processing systems, Safety related communication in open transmission systems, 2001.

[CEN 03] CENELEC – NF EN 50129, Norme européenne, Applications ferroviaires: systèmes de signalisation, de télécommunications et de traitement systèmes électroniques de sécurité pour la signalisation, 2003.

[CEN 11] CENELEC – EN 50128. Railway applications. Communications, signalling and processing systems. Software for railway control and protection systems, January 2011.

[COU 77] COUSOT P., COUSOT R., "Abstract interpretation: a unified lattice model for static analysis of programs by construction or approximation of fixpoints", *Conference Record of the 4th Annual ACM SIGPLAN-SIGACT Symposium on Principles of Programming Languages* (POPL '77), Los Angeles, January 1977, p. 238-52, ACM Press, New York, 1977.

[COU 00] COUSOT P., "Interprétation abstraite", *TSI*, vol. 19, no. 1-3, www.di.ens.fr/~cousot/COUSOTpapers/TSI00.shtml, 2000.

[DIJ 76] DIJKSTRA E.W., *A Discipline of Programming*, Prentice Hall, Upper Saddle River, 1976.

[DOR 08] DORMOY F.-X., "Scade 6, a model based solution for safety critical software development", *Embedded Real-Time Systems Conference*, 2008.

[HAD 06] HADDAD S., KORDON F., PETRUCCI L. (ed.), *Méthodes formelles pour les systèmes répartis et coopératifs*, Hermès-Lavoisier, Paris, 2006.

[HAL 91] HALBWACHS N., CASPI P., RAYMOND P., PILAUD D., "The synchronous dataflow programming language Lustre", *Proceedings of the IEEE*, vol. 79, no. 9, p. 1305-1320, September 1991.

[HAL 05] HALBWACHS N., "A synchronous language at work: the story of Lustre, MEMOCODE '05", *Proceedings of the 2nd ACM/IEEE International Conference on Formal Methods and Models for Co-Design*, 2005.

[HAT 94] HATTON L., SAFER C., *Developing Software for High-Integrity and Safety-Critical Systems*, MCGraw-Hill, New York, 1994.

[HOA 69] HOARE CAR, "An axiomatic basis for computer programming", *Communications of the ACM*, vol. 12, no. 10, p. 576-583, 1969.

[HUL 05] HULL E., JACKSON K., DICK J., *Requirements Engineering*, Springer, New York, 2005.

[IDA 06] IDANI A., B/UML: mise en relation de spécifications B et de descriptions UML pour l'aide à la validation externe de développements formels en B, PhD thesis, Joseph Fourier University, Grenoble, November 2006.

[IDA 07a] IDANI A., BOULANGER J.-L. PHILIPPE L., "A generic process and its tool support towards combining UML and B for safety critical systems", *CAINE 2007*, San Francisco, 7-9 November 2007.

[IDA 07b] IDANI A., OKASA OSSAMI D-D., BOULANGER J.-L., "Commandments of UML for safety", *2nd International Conference on Software Engineering Advances IEEE CS Press*, August 2007.

[IDA 09] IDANI A., BOULANGER J.-L., PHILIPPE L., "Linking paradigms in safety critical systems", *Revue ICSA*, September 2009.

[IEC 98] IEC 61508. Sécurité fonctionnelle des systèmes électriques électroniques programmables relatifs à la sécurité, Norme internationale, 1998.

[IEC 05] IEC 61511, Sécurité fonctionnelle – Système instrumenté de sécurité pour le secteur des industries de transformation, Norme européenne, March 2005.

[IEC 06] IEC 60880, Centrales nucléaires de puissance – instrumentation et contrôles commande importants pour la sécurité, Aspects logiciels des systèmes programmés réalisant des fonctions de catégories A. Norme internationale, 2006.

[ISA 05] ISA, Guide d'interprétation et d'application de la norme IEC 61508 et des normes dérivées IEC 61511 (ISA S84.01) et IEC 62061, April 2005.

[ISO 95] ISO/IEC Information technology – Programming languages – Ada, ISO/IEC 8652:1995.

[ISO 99] ISO, ISO C standard 1999. Technical report, www.open-std.org/jtc1/sc22/wg14/www/docs/n1124.pdf, 1999.

[ISO 00] ISO/IEC 15942, Technologies de l'information – Langages de programmation – Guide pour l'emploi du langage de programmation Ada dans les systèmes de haute intégrité, 2000.

[ISO 03] ISO/IEC 14882:2003(E), Programming Languages – C++, American National Standards Institute, New York 10036, 2003.

[ISO 08] ISO 9001:2008. Systèmes de management de la qualité – Exigences, 2008.

[ISO 09] ISO, ISO/CD-26262, Road vehicles – Functional safety – non publié, 2009.

[JON 90] JONES C.B., *Systematic Software Development using VDM*, Prentice Hall International, 2nd edition, 1990.

[KER 88] KERNIGHAN B.W., RITCHIE D.M., *The C Programming Language*, Prentice Hall, Upper Saddle River, 1988 (2nd edition).

[LER 09] LEROY X., "Formal verification of a realistic compiler", Communication of ACM, vol. 52, no. 7, p. 107-115, July 2009.

[LIS 90] LISSANDRE M., *Maîtriser SADT*, Armand Colin, Paris,1990.

[LOC 05] LOCKHEED M., Joint strike fighter air vehicle C++ Coding standards for the system development and demonstration program, document no. 2RDU00001, rev.C, December 2005.

[MAT 98] MATRA/RATP, "Naissance d'un Métro. Sur la nouvelle ligne 14, les rames METEOR entrent en scène. PARIS découvre son premier métro automatique", *La vie du Rail et des transports*, no. 1076, special issue, October 1998.

[MEY 98] MEYERS S., *Effective C++: 50 Specific Ways to Improve Your Programs and Design*, Addison-Wesley, 1998 (2nd edition).

[MIS 98] MISRA, MISRA-C: 1998 guidelines for the use of the C language in vehicle based software, Motor Industry Research Association, April 1998.

[MIS 04] MISRA, MISRA-C: 2004 guidelines for the use of the C language in critical systems, Motor Industry Research Association, October 2004.

[MIS 08] MISRA, MISRA-C++: 2008 guidelines for the use of the C++ language in critical systems, Motor Industry Research Association, June 2008.

[MON 00] MONIN J.-F., *Introduction aux méthodes formelles*, Hermes, Paris, 2000.

[MOT 05] MOTET G., "Vérification de cohérence des modèles UML 2.0", *Première journée thématique Modélisation de Systèmes avec UML, SysML et B-Système*, Association française d'ingénierie système, Toulouse, 2005.

[OKA 07] OKALAS O., MOTA J.-M., THIRY L., PERRONNE J.-M., BOULANGER J.-L., "A method to model guidelines for developing railway safety-critical systems with UML", *ICSOFT '07, International Conference on Software and Data Technologies*, Barcelone, 2007.

[OMG 06a] OMG, Unified Modeling Language: Superstructure, version 2.1, OMG document ptc/06-01-02, January 2006.

[OMG 06b] OMG, Unified Modeling Language: Infrastructure, version 2.0, OMG document formal/05-07-05, March 2006.

[OMG 07] OMG, Unified Modeling Language (uml), version 2.1.1, 2007.

[OOT 04] OOTIA, Handbook for Object-Oriented Technology in Aviation (OOTiA), vol. 1,2,3,4: Handbook Overview, Revision 0, 26 October 2004.

[RAM 09] RAMACHANDRAN M., ATEM DE CARVALHO A. (ed.), *Software Engineering and Productivity Technologies: Implications of Globalization, Handbook of Research*, August 2009.

[RAM 11] RAMACHANDRAN M. (ed.), *Knowledge Engineering for Software Development Life Cycles: Support Technologies and Applications*, Leeds Metropolitan University, UK, April 2011.

[RIC 94] RICHARD-FOY M., LEGOFF G., "On-board with safety critical software: implementing safety critical software for high-speed railway transportation", *Alsys World Dialogue*, vol. 8, no. 2, 1994.

[SMI 07] SMITH DAVID J. and SIMPSON KENNETH G. L., *Functional Safety, A Straightforward Guide to Applying IEC 61508 and Related Standards*, second edition, Elsevier, 2007.

[SUT 05] SUTTER H., ALEXANDRESCU A., C++ Coding Standards, 101 Rules, Guideline, and Best Practices, Addison Wesley, Boston, 2005.

[RAS 08] RASSE A., BOULANGER J.-L., MARIANO G., THIRY L., PERRONNE J.-M., "Approche orientée modèles pour une conception UML certifiée des systèmes logiciels critiques", *CIFA Conférence internationale Francophone d'Automatique*, Bucarest, November 2008.

[SPI 89] SPIVEY J.M., *The Z notation - A Reference Manual*, Prentice Hall International, 1989

Glossary

AADL	Architecture Analysis and Design Language
ACATS	Ada Conformity Assessment Test Suite
ACU	Alarm Control Unit
AFIS	French: *Association française d'ingénierie système*; English: French Association of Systems Engineering
AMN	Abstract Machine Notation
ANSI	American National Standards Institute
APU	Auxiliary Power Unit
ASA	Automata and Structured Analysis
ASIL	Automotive SIL
ATO	Automatic Train Operation
ATP	Automatic Train Protection
ATS	Automatic Train Supervision
BDD	Binary Decision Diagram
CAN	Controller Area Network
CAS	Computer-Assisted Specification
CBTC	Communication-Based Train Control
CbyC	Correctness by Construction

CCP	Centralized Control Point
CdCF	French: *cahier des charges fonctionnel*; English: functional requirements specification
CENELEC[1]	European Committee for Electrotechnical Standardization
CMMI	Capability Maturity Model Integration
CPU	Central Processor Unit
DAL	Design Assurance Level
DoD	Department of Defense
DV	Design Verifier
E/E/PE	Electric/Electronic/Programmable Electronic
EAL	Evaluation Assurance Level
ECU	Electronic Control Units
ERTMS	European Rail Traffic Management System
FADEC	Full Authority Digital Engine Control
FBD	Function Block Diagram
FC	Failure Condition
FDA	Food and Drug Administration
FMECA	Failure Mode and Effects Criticality Analysis
FT	Functional Testing
GPS	GNAT Programming Studio
GSL	Generalized Substitution Language
GUI	Graphical User Interface
HIL	Hardware In the Loop
HR	Highly Recommended
IDE	Integrated Development Environment

1 See www.cenelec.eu.

IEC[2]	International Electrotechnical Commission
iFACTS	interim Future Area Controls Tools Support
IL	Instruction List
IMAG	French: *Institut de mathématiques appliquées de Grenoble*; English: Institute of Applied Mathematics of Grenoble
IPSN	French: *Institut de protection et de sûreté nucléaire*; English: Institute of Nuclear Protection and Safety
IRSN	French: *Institut de radioprotection et de sûreté nucléaire*; English: Radioprotection and Nuclear Safety Institute
ISO[3]	International Organization for Standardization
ISSE	International Symposium on Software Reliability Engineering
IT	Integration Testing
KCG	Qualifiable Code Generator
KLOC	1000 LOC
LaBRI	French: *Laboratoire bordelais de recherche en informatique*; English: Bordeaux Laboratory for Computer Research
LD	Ladder Diagram
LOC	Lines of Code
MaTeLo	Markov Test Logic
MBD	Model-Based Design
MBT	Model-Based Testing
METEOR	*METro Est Ouest Rapide*, train operation system used by the Paris metro
MISRA[4]	Motor Industry Software Reliability Association
MMI	Man-Machine Interface

2 See www.iec.ch.
3 See www.iso.org/iso/home.htm.
4 See www.misra.org.uk.

MPU	Main Processor Unit
MTC	Model Test Coverage
MTTF	Mean Time To Failure
MU	Multiple Unit
NHMO	NATO HAWK Management Office
NR	Not Recommended
NSA	US National Security Agency
NSE	No Safety Effect
OCR	Optical Character Recognition
OFP	Operational Flight Plan
OMG[5]	Object Management Group
OOTIA[6]	Object-Oriented Technology in Aviation
OPRI	French: *Office de protection contre les rayonnements ioisants*; English: Office of Ionising Radiation Protection
OS	Operating System
PADS	French: *pilote automatique double sens*; English: two-way autopilot
PAI-NG	French: *poste d'aiguillage informatisé de nouvelle génération*; English: next-generation computerized signaling control
PO	Proof Obligation
PTS	Problem of the Traveling Salesman
PWM	Pulse With Modulation
QTP	Quick Test Professional
R	Recommended
R&D	Research and Development

5 See www.omg.org.

6 See www.faa.gov/aircraft/air_cert/design_approvals/air_software/oot.

RAMS	Reliability, Availability, Maintainability, and Safety
RAT	Ram Air Turbine
RATP[7]	French: *Régie autonome des transports parisiens*; English: Autonomous Operator of Parisian Transports
RER	French: *Réseau express régional*; English: Regional Express train
RFT	Rational Functional Tester
RM	Requirement Management
ROM	Read-Only Memory
SACEM	French: *Système d'aide à la conduite, à l'exploitation et à la maintenance*; English: Assisted Driving, Control, and Maintenance System
SADT	Structured Analysis and Design Technique
SAET	French: *Système d'automatisation de l'exploitation des trains*; English: Automation System of Train Operations
SAO	French: *Spécification assistée par ordinateur*; English: Computer-Aided Design
SAS	Software Architecture & design
SCADE	Safety-Critical Application Development Environment
SFC	Sequential Function Chart
SHOLIS	Ship/Helicopter Operational Limits Instrumentation System
SIL	Safety Integrity Level
SIS	Safety Instrumented System
SMDS	Software Module Design Specification
SMTP	Software Module Test Plan
SOC	System On a Chip
SPICE	Software Process Improvement and Capability Determination

7 See www.ratf.fr.

SPIN	French: *système de protection intégré numérique*; English: Digital Integrated Protection System
SRS	Software Requirement Specification
SSIL	Software SIL
ST	Structured Text
SU	Single Unit
SUT	Software Unit Test
TCMS	Train Control Management System
TCO	French: *tableau de contrôle optique*; English: Visual Control Panel
TFTA	Terrain Following Terrain
TIS	Tokeneer ID Station
TOR	French: *Tout ou Rien*; English: Hit-or-Miss
UML	Unified Modeling Language
UT	Unit Testing
V&V	Verification and Validation
VC	Verification Condition
WCET	Worst-Case Execution Time
WP	Weakest Precondition

List of Authors

Pierre BIEBER
ONERA
Toulouse
France

Jean-Louis BOULANGER
CERTIFER
Anzin
France

Michael BUTLER
University of Southampton
UK

Judith CARLTON
Escher Technologies Ltd
Aldershot
UK

David CROCKER
Escher Technologies Ltd
Aldershot
UK

Anthony FAUCOGNEY
ALL4TEC
Massy
France

Héléne LE GUEN
ALL4TEC
Laval
France

Patrick MUNIER
MathWorks SAS
Paris
France

Ian O'NEILL
Altran Praxis Ltd
Bath
UK

Aryldo G. RUSSO Jr.
Aes Group
São Paulo Area
Brazil

Asieh SALEHI FATHABADI
University of Southampton
UK

Christel SEGUIN
ONERA
Toulouse
France

Renato SILVA
University of Southampton
UK

Frédérique VALLÉE
ALL4TEC
Massy
France

Index

R

S

T

V

W